Nanotechnology and Sustainable Development

Routledge Studies in Technology, Work and Organization

1 **Information Society and the Workplace**
Edited by Tuula Heiskanen and Jeff Hearn

2 **Technological Change and Organizational Action**
Edited by David Preece and Juha Laurila

3 **Managing Complex Projects**
Networks, Knowledge and Innovation
Neil Alderman, Chris Ivory, Ian McLouglin, Alfred Thwaites and Roger Vaughan

4 **Information and Communication Technologies in Rural Society**
Being Rural in a Digital Age
Grete Rusten and Sarah Skerratt

5 **Software and Organizations**
The Biography of the Enterprise-wide System or How SAP Conquered the World
Neil Pollock and Robin Williams

6 **Technological Communities and Networks**
Triggers and Drivers for Innovation
Dimitris Assimakopoulos

7 **Health Technology Development and Use**
From Practice-Bound Imagination to Evolving Impacts
Sampsa Hyysalo

8 **Nanotechnology and Sustainable Development**
Claire Auplat

Nanotechnology and Sustainable Development

Claire Auplat

Routledge
Taylor & Francis Group
LONDON AND NEW YORK

First published 2012
by Routledge
2 Park Square, Milton Park, Abingdon, Oxon OX14 4RN
52 Vanderbilt Avenue, New York, NY 10017

Routledge is an imprint of the Taylor & Francis Group, an informa business

Typeset in Sabon by IBT Global.

Library of Congress Cataloging-in-Publication Data
Auplat, Claire.

Nanotechnology and sustainable development / by Claire Auplat.
p. cm. — (Routledge studies in technology, work, and organizations ; 8)
Includes bibliographical references and index.
1. Nanotechnology—Environmental aspects. 2. Information technology—Environmental aspects. 3. Sustainable development. I. Title.
T174.7.A967 2011
338.9'27—dc23
2011021855

ISBN13: 978-0-415-89182-0 (hbk)
ISBN13: 978-0-203-15581-3 (ebk)

Contents

APPENDICES

Figures

Acknowledgments

This book could only see the light of day thanks to a large number of people. First, I would like to thank the numerous persons who kindly provided me with information and raw material, whether in the form of interviews, documents, or advice for further encounters. From nanotechnology research facilities at Rice University (Wade Adams) to the Nano and Converging Sciences and Technologies Unit at the European Commission in Brussels (Renzo Tomellini), to the functioning of an NGO (nongovernmental organization) in Winnipeg, Canada (Pat Mooney) or to the understanding of the motivations of early nanotechnology enterprises in Lyon, France (Patrick Valade, Babolat), I owe the 'feel' of my research to these people, to the time they spared for me, to the contacts they facilitated, and to the openness of our discussions. I cannot list them all here, but I would like to express my deepest gratitude to them for their time and support.

David Gann and Nicos Nicolaou played a very important role in introducing me to entrepreneurship research. As a specialist of entrepreneurship in high technology ventures, Nicos steered me through the understanding of some of its specific mechanisms, and our discussions were extremely helpful.

I had worked on sustainable development issues previously, particularly through my early involvement with the World Forum on Sustainable Development (FMDD) hosted at the French Senate, but my initial acquaintance with the environmental issues linked to the development of nanotechnologies took place at the Center for Biological and Environmental Nanotechnology (CBEN) in Houston, and I am especially grateful to its executive director Kristen Kulinowski for her early support, as well as for her help in the constitution of part of my database.

I received very helpful feedback on this work during the course of several conference presentations, including the EURAM conference in Oslo in 2006, the EGOS conference in Vienna in 2007, the Cambridge Conference on Regulation, Inspection & Improvement in Cambridge (UK) in 2007, the Nanotechnology and Occupational Health and Safety Conference in Santa Barbara in 2008, and MINATEC Crossroads 2010. I was extremely fortunate to discuss particular points of my research with specialists of very different fields such as Alain Fayolle, Emmanuel Flahaut, Richard B. Freeman, Michael Lounsbury, Nelson Phillips, Scott Shane, Mark Welland, Mark Wiesner, and Lynne Zucker, among others. I would like to extend my deepest thanks to all for their contributions, some of which have been pivotal. My students at Sciences Po also deserve credit for adding new perspectives, especially on environmental issues, and I thank them greatly for their insightful comments. Finally, I give my deepest thanks to Routledge, and specifically to Terry Clague, David Preece, Laura Stearns, Stacy Noto, and Michael Watters for their encouragements, for trusting me with this project, and for all their warm and efficient support.

Introduction

This book explores the development of nanotechnologies from research funding to market stage. It studies the emergence of new institutions or the modification of existing ones, and it seeks to account for the relations between the two. Its purpose is to show that the institutional changes that coexist with the development of nanotechnologies are a sign of a deep societal transformation and that this deep societal transformation is based on an increasing awareness of the importance of the concept of sustainable development.

According to Feenberg and Hannay (1995), technology is one of the major sources of public power in modern societies, and our daily lives are affected by the fact that political democracy is largely overshadowed by the enormous power wielded by the masters of technical systems. For example, corporate and military leaders and professional associations of groups such as physicians and engineers have more control over the selection of innovations—and therefore over our lifestyles, from patterns of urban growth and communication systems to our experiences as patients and consumers—than do all the governmental institutions of our society put together.

With the rise of nanotechnologies, however, this model is changing: a new form of democracy is taking shape *which is forcing itself into* the process of technological emergence. This book is designed to show how this process is taking place, how it relates to institutional change, and, finally, why it is happening.

The study of institutional emergence is a new discipline. Its roots can be found in new institutional theory, itself a fairly recent field of research. The conceptual foundations of modern organizational institutionalism were established in the works of Meyer and Rowan (1977), Zucker (1977), DiMaggio and Powell (1983, 1991), and Meyer and Scott (1983). The birth of new institutions, the way they take shape, and the shape they actually take are fascinating topics of research, even if the complexity of the interactions at play may be sometimes rather dispiriting. The empirical field of nanotechnologies constitutes an excellent domain for finding clues about institutional processes. Indeed, the technology is the nexus of a whole range of technical, economic, and social activities as well as the source of much debate relating to ethics, regulation, economics, and politics.

New institutional theory claims that institutions operate in an environment consisting of other institutions that influence them. The interactions among institutions create a general framework that shapes organizations. Then, even though organizations evolved very differently, they usually end up having a similar structure because they need to establish legitimacy within the world of institutions. This process of alignment through mimicry was called by neo-institutionalists "institutional isomorphism" (DiMaggio and Powell 1983).

In a sense, the establishment of legitimacy is at the core of this research. Indeed, if institutions are providers of organizational legitimacy, understanding the mechanisms of institutional change around the emergence of nanotechnologies is central to understanding their future development. In their work on institutional entrepreneurship, Lawrence, Hardy, and Phillips (2002) have analyzed the process through which organized actors—the institutional entrepreneurs—structure the institutional contexts in which they operate in order to create opportunities to generate new institutions. Maguire, Hardy, and Lawrence (2004) have for their part tried to identify the specifics of institutional entrepreneurship in emerging fields. Building on extensive previous research (DiMaggio 1988; Fligstein 1997; Rao, Morrill, and Zald 2000; Seo and Creed 2002), they show that institutional entrepreneurship results from relationships between interests, agencies, and institutions, and that the way in which institutional entrepreneurs connect their change projects to the activities and interests of other actors in a field and craft their projects to fit the conditions of the field itself is at the heart of their success. I draw from their concurring conclusions to provide the conceptual framework for this work.

This book provides an integrated analysis of the combined actions of stakeholders who act as institutional entrepreneurs to demonstrate that institutional emergence in nanotechnologies marks a rupture from previous forms of institutional emergence. It establishes that it is not one group of stakeholders, one set of actions alone that can bring about this rupture, but a complex web of interactions. It also reveals that the connecting element between all these interactions, the source of the rupture from previous forms of institutional emergence, is increasing awareness of the importance of sustainable development.

In their groundbreaking work, Callon and Latour explained that the institutionalization of science and technology rests on a chronological process that includes several steps (Latour 1988; Callon, Law, and Rip 1986). Science and technology first emerge as disconnected facts and gradually stabilize and acquire ontology through the settlement of controversies concerning their validity. For the researchers, it is not an intrinsic natural value but the settlement of *controversy* that lies at the root of scientific institutional emergence. Controversy is followed by *translation*, an operation that creates a common language of reference so that previously unrelated entities can be connected and seen in interaction. The process of translation gives unity and intelligibility to otherwise obscure and apparently disconnected entities. It allows the institutionalization process to move into the domain of public understanding. It is also the lifeblood of the network, an ensemble of meta-organizations including human and non-human *actants* that interact together to form the new institutions.

Two new elements have emerged since Callon et al. developed their theory. One is a global awakening to the importance of sustainable development; the other is the widespread use of information technology and espeically the internet. Although the process of institutionalization still follows the same stages, this book shows how these two elements have modified their model.

What about sustainable development? One cannot hope to understand the concept without looking at the historical background. *Development* and *sustainability* are two terms that appear incompatible in the face of it. They have spawned a huge literature that has produced so many theories and counter theories that one could safely say that they have come to underpin a large part of economic thinking today. The words themselves appeared for the first time in the outcomes of a study commissioned by the United Nations in 1972 (UN 1972). The notion had emerged earlier, but it took some twenty years to shape as a collective preoccupation. To understand the mechanism of its emergence, one has to go back in history.

Thomas R. Malthus noticed in the eighteenth century that the power of population was far superior to the power of the earth to produce subsistence for human beings, and that it was only because there were epidemics, wars, and various exogenous causes of extermination that population could be kept in check at a sustainable level. He then calculated what would happen if the earth's population was not decimated at regular intervals as it always had been. His calculation led him to predict in 1798 that population would one day—which he situated in the middle of the nineteenth century—outrun food supplies (Malthus 1999). Although this prediction was proved wrong, essentially because of flaws in his methodology, his theory became extremely influential and can be seen as having introduced the concept of the necessity for sustainability in human development.

A century later, the Swedish scientist Svante Arrhenius—who received a Nobel Prize in 1903 for his work on electrolysis—calculated that if the volume of carbon dioxide present in the atmosphere were to double this would lead to a temperature rise of the planet of 4 degrees by the end of the twentieth century. He predicted that the intensive use of fossil fuels would warm the earth's climate. His study stayed untouched for many years because scientists lacked the tools to confirm or refute his intuitions. It was only in the 1960s with the development of the first computers that people could start collecting data and modeling airflows to improve weather forecasting. Until then, climate research only relied on very basic observations and technologies. The first experiments led to the discovery of the thinning of the ozone layer and of the nature of acid rain, and also led to a fear of a general pollution of the higher atmosphere due to the growing number of supersonic planes. In 1965, three groups of American researchers (Geophysical Fluid Dynamics Laboratory, UCLA Department of Meteorology, and the National Center for Atmospheric Research) began to draw the first lessons from their attempts to model airflows (Weart 2003).

The next landmark contribution to the emergence of the concept of the environment not as an exogenous entity but as a component of the evolution of the human species came from the work of Donella Meadows, an American researcher who had begun studying chemistry and biophysics at Harvard before joining Jay Forrester's team at MIT (Massachusetts Institute of Technology). Forrester (1961) was at the forefront of systems dynamics, and the research group developed a computer model, "World3," which was a computer simulation of the interactions among population growth, industrial growth, food production, and the limits to the earth's ecosystem. The book that followed this research, The Limits to Growth (Meadows and Meadows 1972), used the computer model to show that over the course of one century, unlimited growth would lead to a collapse of world resources and that if current trends were not modified, the world-system would collapse around 2000. The report was very controversial, but it made it clear for the first time that there were physical limits to the world's resources, and that human development might be restrained by the fact that the earth's resources were limited.

This period of time was also the beginning of space missions. Thanks to the pictures taken by NASA in December 1968 from the *Apollo 8* orbiter, people could see the earth from the outside for the first time in history. It made them realize that it was a small ball in the immensity of space, and that it was both very beautiful and very fragile.

This brief chronology of the emergence of the concept of sustainable development shows that technological progress was at the origin of people's awareness of the complexity and vulnerability of the environment. This awareness was double-sided: on the one hand, people started realizing that through their activities they were altering the conditions that made their life on Earth possible. On the other hand, they discovered that the mechanisms of interactions between human activities and the environment were extremely complex and interconnected: air pollution in eastern Europe resulted in acid rain in Sweden (Oden 1968); discrete problems of pollution such as the persistence and spread of DDT or oil spills in the oceans led to changes in the equilibrium of ecosystems (SCEP and Matthews 1970); human activities could result in climate change (SMIC 1971).

The rest of the story is well known. The phrase "sustainable development" appeared for the first time in writing in 1980 in a report called *World Conservation Strategy: Living Resource Conservation for Sustainable Development*, which was co produced by IUCN (International Union for the Conservation of Nature), UNEP (United Nations Environmental Program), and WWF (World Wildlife Fund) (IUCN, UNEP, and WWF 1980). Then,

UNEP published in 1987 a report resulting from a three-year study: *Our Common Future* (Brundtland and World Commission on Environment and Development 1987). The report, also called the Brundtland report, defined sustainable development as "development that meets the needs of the present without compromising the ability of future generations to meet their own needs."[1] In 1991, the successor to the World Conservation Strategy, IUCN, produced with UNEP and WWF a new report called *Caring for the Earth: A Strategy for Sustainable Living.* In the report, sustainable development received a new definition: "Improving the quality of human life while living within the carrying capacity of supporting ecosystems."[2] The report revealed the need to implement new strategies to allow continuing development while preserving the world's capacity to preserve and regenerate itself. This new definition integrated economic and social development with the preservation of the environment and of natural resources.

The notion of sustainable development has now become deeply ingrained in society, and the reason it has done so in such a radical way is due to the widespread use of information technology. First, the period we are studying, between the 1970s and the end of the first decade of the twenty-first century, saw the generalization of television. This allowed pictures such as *Apollo 8*'s "Earth Rise" to be seen by a very large number of people on Earth. These people could see and feel for themselves the beauty and vulnerability of the planet. In addition, it so happened that the release of the Brundtland report *Our Common Future* coincided with human and environmental tragedies that were widely broadcast on the news. The report itself drew a list of these disasters in the form of a chart. In the course of the three years preceding the 1987 publication, there had been a drought in Africa that had led to the death of approximately one million people; a gas leak at the Bhopal pesticide factory in India had killed more than two thousand people and injured over two hundred thousand more; the Chernobyl nuclear reactor explosion in Ukraine, then part of the Soviet Union, had sent nuclear fallout across Europe, increasing the risks of future human cancers; liquid gas tanks had exploded in Mexico City, killing one thousand and leaving thousands more homeless; agricultural chemicals, solvents, and mercury had flowed into the Rhine River during a warehouse fire in Switzerland, killing millions of fish and threatening drinking water in Germany and the Netherlands; and an estimated sixty million people—most of them children—had died of diarrheal diseases related to unsafe drinking water and malnutrition.[3] Because of the broadcasting of all these tragedies, many people realized that the planet was small and easily damaged, and some found connections between technological progress and disaster. So the globalization of information contributed to increase awareness of environmental issues.

Then, the development of the Internet took the traveling of information and the popularization of science even further to the stage we now know and take for granted. It was at this stage that nanotechnology came to be heard of, and its emergence as a field coincided historically with a growing awareness of the importance of sustainable development—both for citizens and for the economy. It is the main reason why, as I explain in this book, institutional emergence in nanotechnologies marks a rupture from previous forms of institutional emergence: nanotechnologies arrived at a point in time when technological progress and social evolution reached a point of convergence with an awareness of the importance of sustainable development.

Part I

Nano, the Next Dimension of Sustainable Development

1 Background and Road Map

A search of the literature shows that, in fact, there has been a time lag for the acknowledgement of sustainable development as a matter of research between social scientists and economists (Figure 1.1). Sustainable development was a source of interest for social scientists—and for a minority of concerned scientists—already in the mid-1980s (Stiglitz 1998), whereas it only started becoming one for economists around 2000.

Books on the practical implementation of sustainable development in business began being published in the early 2000s (Hitchcock and Willard 2006). Yet a decade later, some experts and scholars still denied its real impact on business activity. Although a growing number of listed companies included a chapter on sustainable development in their business reports—companies listed on the French stock exchange were obliged by law to include social and environmental information in their annual reports, as an example (European Commission 2007)—these specialists considered that there were wide gaps between rhetoric and reality and that sustainable development related more to communication processes than to any deep change in business models (Veleva 2007). However, recent research points to the contrary. It indicates that sustainable development is gradually changing the focal point of some segments of business activity and that this is the premise of a more general change in business models. In 2008, the French consultancy RSE paradigm21 carried out a study to explore the operationalization of sustainable development in firms (Rios 2008). It drew its conclusions from an empirical study bearing on some forty multinational companies in sectors as diverse as food, telecommunications, retailing, electronics, cars, transport, and banking.

An important finding was that companies were then producing the second generation of sustainable development managers, and this second generation was extremely different from the first one. Whereas the first generation might have been constituted of people coming from various backgrounds with sometimes no clear idea of their role and aims, the second one was often composed of managers who had a specific agenda. The first generation was there mostly because there was a general feeling that integrating sustainable development in organizations was the thing to do since everyone else

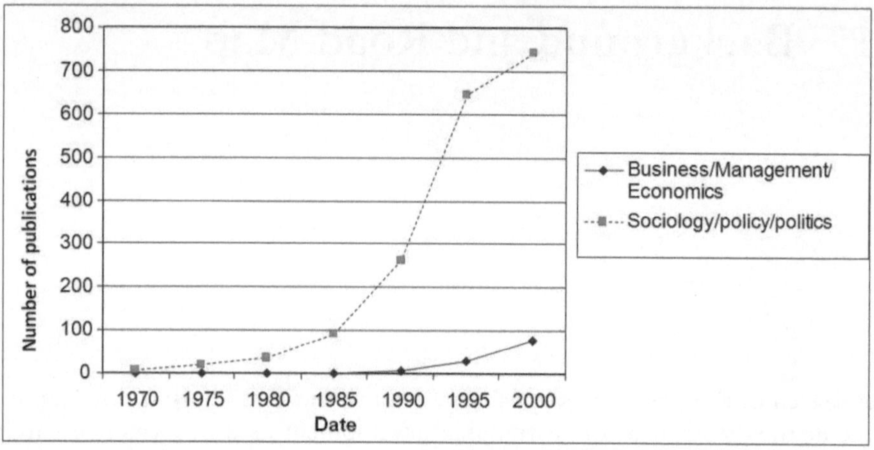

Figure 1.1 Number of publications related to sustainable development in business/management/economics literature and in sociology/policy/politics literature.
Source: EBSCO, JStor, author's compilation.

was doing the same. At that time, integrating sustainable development in organizations usually amounted to adding paragraphs on sustainable development in business reports or creating positions of sustainable development managers. However, this first generation of sustainable development managers demonstrated through original initiatives the idea that constraining environmental regulation could also bring about growth opportunities. At the same time, training became widely available, beginning with courses on sustainable development in the most prestigious universities. Far from considering sustainable development as a form of window dressing, the second generation of sustainable development managers often saw it as a source of new business development. Consequently, sustainable development gradually began bringing about shifts in business strategies.

The RSE paradigm21 study reflected that there were differences in the way sustainable development was integrated in business models in different sectors. It highlighted two trends. The first trend consisted of adding new functionalities to existing products or services so that they complied with sustainable development targets. This trend did not signal fundamental change in business models. However, the second trend was completely different, and this is what is important to keep in mind to understand the coindustrial and institutional emergence of nanotechnologies. By actively seeking to meet environmental targets, some companies reinvented the way they worked. This involved strong, long-term support from the top management, with a new understanding of performance and new forms of interactions between business functions such as R&D, resource acquisition, human resources, marketing, or public relations. But it actually led to shifts in R&D focus as well as in development strategies.

A few examples will help demonstrate the change. While only 2 percent of the world's wastewater was recycled, the Company Degrémont, a subsidiary of Suez Environnement specializing in water treatment, launched a new program to build factories treating wastewater so that it could safely be used in agriculture. The first such factory—treating 4000 liters of wastewater per second—was built in Milan, Italy, in 2004. In 2010, it handled wastewater for a population equivalent of 1,050,000 and was capable of accommodating exceptional storm water loads. Practically all of the treated water could be reused for farm irrigation, an important part of the economy of the region.

The French national rail company SNCF (Société Nationale de Chemins de Fer) began publishing an "eco-comparator" in 2006. The EcoComparateur was designed in collaboration with ADEME (Environment and Energy Management Agency). It enabled people to compare different modes of transportation to a given destination thanks to an environmental index using CO_2 kg as a unit. Its impact on customers was rapid, with the website receiving over 850,000 visits in two years. Soon technical innovation and innovation in management practices became major drives to improve the score of the train as a means of transport on the EcoComparateur. This means that the publication of the EcoComparateur led to changes in the SNCF business strategy. In 2008, SNCF signed a framework agreement with ADEME to improve its environmental practices. This included, among other things, change in the management of refuse (quantity and type of refuse left onboard trains) and therefore change in the type of food and packaging used by train restaurants. It also involved changes to improve energy efficiency (lighter components for lighter trains, experiments with alternative fuels, different rail systems, etc.). Another sector that was undergoing big changes because of the awareness of sustainable development was the building environment. According to Lafarge, a world leader in building materials, the challenges of sustainable construction concerned all players in the building sector. Lafarge worked closely with industrial associations, energy suppliers, and architects to identify building methods that were cleaner and more environmentally friendly. In 2006, within the framework of the World Business Council for Sustainable Development (WBCSD), Lafarge initiated the Energy Efficiency in Buildings (EEB) project in partnership with United Technologies Corp., the world's leading supplier of capital goods. One of the major goals of the EEB project was to construct buildings that were self-sufficient in terms of energy and that did not emit any CO_2. In just a few months, nine multinational companies including DuPont, Gaz de France, Cemex, Tokyo Electric Power Company, and ITT joined the project. These examples illustrate that the concept of sustainable development brought deep strategic changes not only inside companies, but also in the way they related to others and worked together on shared projects. Change affected companies internally and externally, in R&D

orientations as well as in overall business strategies. Sustainable development is gradually becoming a full component of economic activity, and the time has come to explore its interactions with organizations and management. So, what about nanotechnologies? How do they connect to sustainable development issues?

If we continue with the areas of human activity just discussed, among the many promises of nanotechnologies are the ability to eliminate waste and toxins from production processes, to remove contaminants from water, and to improve fuel efficiency. The report of the 2007 Workshop on Nanotechnology for Cement and Concrete sponsored by The National Concrete Pavement Technology Center and the National Science Foundation opened the door to active research into the potential of nanotechnology applications in cement and concrete for the building environment in particular, and for the environment in general (Taylor et al. 2007). Nanotechnology, then, appears like a cross sector technology that may bring new solutions to many areas that are concerened by issues of sustainable development. At the same time, researchers and environmentalists have long cautioned that working with molecular-scale materials offers unforeseen threats to humans, animals, and the planet. So nanotechnologies themselves can be seen either as a solution to sustainable development issues or as a source of problems to come. This was made clear by the UNEP (United Nations Environmental Program) in its 2010 report. In the section entitled "Harmful Substances and Hazardous Waste," the report mentioned that continuing concerns and unanswered questions remained about nanomaterials, notably as a result of the challenges stemming from how to assess their hazards and risks and regulate them (UNEP 2010). But what is so special about nanotechnology? And what, exactly, are nanotechnologies?

Nanotechnology is a generic word that refers to science, technologies, products and processes involving the manipulation of matter at the level of atoms and molecules, at the scale of the nanometer (nm)—that is, of a billionth of a meter (10^{-9} m)—in order to obtain new and unique properties not found in the bulk forms of the particles. It was the lecture that the American Nobel Physics Prize recipient Richard Feynman gave in 1959 at the annual meeting of the American Physical Society at the California Institute of Technology (Caltech) that made people realize that the characteristics of matter could be changed by acting on its constituent atoms. In this lecture "There Is Plenty of Space at the Bottom," he envisaged for the first time what this could mean:

> But I am not afraid to consider the final question as to whether, ultimately—in the great future—we can arrange the atoms the way we want; the very *atoms*, all the way down! What would happen if we could arrange the atoms one by one the way we want them.
>
> (Feynman 1960)

The word itself was used for the first time in 1974 by Norio Taniguchi, a scientist from Tokyo University (Taniguchi 1974). The concept of nanotechnology refers to working on matter at the atomic and molecular scale: by modifying the ways atoms are placed or grouped together, people can either obtain known products with new characteristics, or create completely new matter, whether living or inert. Although not everyone agrees on this as a criterion, nanotechnology tends to refer as a rule to structures that have at least one dimension that is smaller than 100 nm. Even if people started conceiving the immense potential of manipulating matter by acting on its basic components in the 1950s, one had to wait until the 1990s to be technically able to see and to work with such small-scale elements. Today, nanotechnologies cover very diverse fields, which can be gathered under two major areas of development: first, all the computing and optical tools that make it possible to visualize nanometric components and to work at the nanoscopic scale; and second, all the fabrication processes involving nanoparticles.

To manipulate the atoms that constitute any form of matter, there are two opposite approaches that can be taken: one can either start from the bottom or from the top.

STARTING FROM THE BOTTOM

With this method, the idea is to start from the atoms themselves and to assemble them or make them assemble according to a very specific design, in a way that could remind one of children using LEGO blocks to make their own constructions. In 1989, scientists from the IBM lab at the Almaden Research Center in San José, California, placed 35 xenon atoms on the surface of a nickel crystal to form the letters IBM. To do so they used a scanning tunneling microscope (STM), which allowed them to see and manipulate the atoms one by one. The microscope had been invented in the IBM Zurich lab and brought its inventors a Nobel Prize in Physics in 1986.[1]

However, to this day, manipulating individual atoms remains in the domain of laboratory science. One reason is the price of the equipment, essentially very sophisticated microscopes that can cost well over a hundred thousand euros (€100,000). The other reason is that the time necessary to manipulate the atoms makes it absolutely impossible to assemble enough molecules to create matter in a sizeable way. Richard Smalley, a chemist who received a Nobel Prize for the discovery of a particular type of nanocarbon, made a quick calculation based on the fact that making a mole of something of about one ounce (30 grams) would require at least 6×10^{23} bonds, one for each atom. He explained that at the frenzied rate of creating 10^9 bonds per second, it would take nineteen million years to do so! (Smalley 2001)

Seemingly more promising is manipulating matter from the bottom using molecular self-assembly. Eric Drexler, an engineer best known for popularizing the potential of molecular nanotechnology (MNT) in the 1980s, championed this technique in his book *Engines of Creation: The Coming Era of Nanotechnology* (Drexler 1986). He proposed using nanoscopical self-reproducing robots, which he called 'assemblers,' to build matter starting from atoms. He based his vision on the fact that this is precisely what happens with biological reactions in nature. When cells are created or biological actions take place, matter is being created or modified according to predetermined designs. Living cells, for example, are nanoscale devices that are constantly assembling complex structures one or two molecules at a time. For instance, cells construct proteins by putting a chain of amino acids together in the right order, following the 'construction manual' provided by the DNA. In Drexler's vision, assemblers would guide chemical reactions leading to the construction of new matter by moving reactive molecules in the right places. This technique blurs the divide between biology, chemistry, and physics, and is progressing extremely fast in spite of the early criticisms that Drexler had to face. The major criticism addressed to Drexler was that he presented his vision as if it could come true in the near future, which created a great amount of hype, whereas scientists knew that the current advancement of science made it a dream for a great number of decades and possibly centuries. However, it did make people understand and realize what modifying matter could mean in practical terms, and it did boost research in this direction. (Drexler and Smalley 2003)

Starting from the Top

The second approach to creating nanostructures is to manipulate matter starting from the top, that is, to modify the atomic composition of existing matter through a succession of various operations, usually chemical reactions. Whereas in the bottom-up approach scientists start from atoms and try to force them to assemble in given ways to obtain molecules with specific characteristics that have not been found in nature, in the top-down approach, they start from existing matter and attempt to modify its molecular structure to obtain new structures and products. For instance, carbon is a well-known natural element, which is found in abundance on Earth, and which is present in everything we eat, as well as being a component of our bodies. By submitting carbon-containing molecules to chemical and thermal reactions, scientists are able to create varieties of carbon nanotubes that have never been found in a natural state.[2] These nanoobjects containing only carbon atoms have a tubular form and very interesting properties: they are one hundred times stronger and six times lighter than steel. Some have a diameter that does not exceed one nanometer and can reach 100 000 nm in length. Depending on their morphology, they can look like perfect cylinders, bamboo, or rolls of wrapping paper

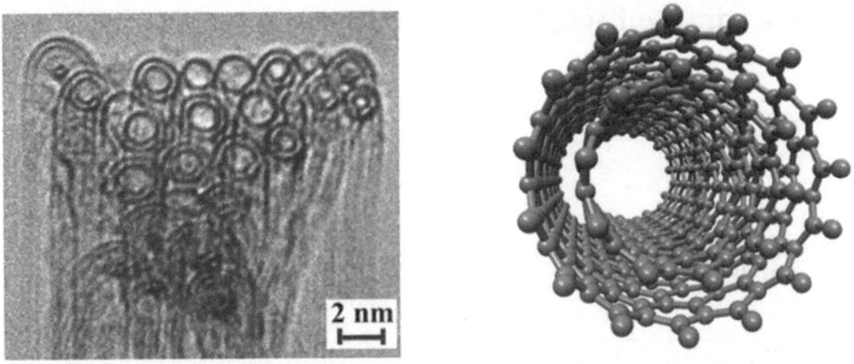

Figure 1.2 Examples of double-walled carbon nanotubes, grouped in a bundle and in perspective.
Source: Flahaut 2003, 2005.

(rare). Depending on their structure, they can be utilized as electric conductors or, conversely, as semiconductors, which makes them extremely interesting in nanoelectronics.

After more than a decade of promise and speculation created by their unique properties, nanotubes are establishing a presence beyond university laboratories and corporate research and development (R&D) centers. Current commercial applications include motor vehicle fuel-system components and specialized sports equipment. In the short term, world demand for nanotubes expanded rapidly from this small base to approximately $250 million in 2010. New functionalized nanotubes applications—whereby specific chemicals are attached to or inserted inside carbon nanotubes to give them additional properties—will come onto the market in the next few years and are expected to increase global revenues to $2.7 billion by 2015. Demand will be driven mainly by the needs of the electronics and data storage, defense, energy, aerospace, and automotive industries[3]. However, a number of issues, including high costs, inadequate purity levels, and insufficient product yields in manufacturing, still need to be addressed.

The French Agency for Environmental and Occupational Health Safety (AFSSET, renamed ANSES in 2010) published reports providing estimates of the most common nanoparticles produced by chemical nature in France, along with their world production estimates (AFSSET 2008, 2010). The top five were carbon, silicium, titanium dioxide, aluminium and rare earths. According to the agency, world production of carbon nanoparticles was ten million tons in 2008, while it was of 2 million tons for silicon and 2 million tons for titanium dioxide. The reports did not quantify the world production of silver nanoparticles but insisted on their important presence in nanobased consumer products in 2010 because of their anti-bacterial properties.

Estimated world production of most common nanoparticles	
Carbon	10,000,000 tons
Silicium	2,000,000 tons
Titanium dioxide	2,000,000 tons
Aluminium	unknown
Rare earth	small quantities

Figure 1.3 Estimated world production of most common nanoparticles.
Source: AFSSET 2008, 2010.

With this background in mind, the main questions that underpin this book are the following:

How do various stakeholders acting as institutional entrepreneurs collectively shape the emergence of nanotechnologies? How does this relate to sustainable development? And how does this coindustrial and institutional emergence mark a rupture from previous mechanisms?

These broad questions cover several subquestions, with the most important ones as follows:

- What makes the nanotechnology context a propitious one for this rupture?
- What are the characteristics of the rupture in institutional emergence?
- What are the effects of this rupture in institutional emergence? (Namely those that are visible, those that are likely, and those that one can only guess at.)

One may wonder whether these are important questions. In fact they are, and the reasons apply at different levels. At the empirical level, nanotechnology is being hailed as one of the most promising sectors of socio- and economic development, whether by public institutions (PCAST 2010), by academic researchers (Chen and Roco 2008; Roco and Bainbridge 2003), or by analysts and investors (Lux Research Inc. 2007). According to the OECD (2009), available forecasts suggest a global market for nanotechnology products in the range of $150–$3,100 billion in 2011 and the creation of some two million new jobs worldwide. Calculations predict that the global market for nanotechnology products will affect the economy in a transversal way in a manner that has not been seen before, and pundits claim that it could bring about a revolution of at least the same magnitude as the Industrial Revolution because of all the changes in our lifestyles that it can bring about. At the same time, many specialists draw attention to the challenges that such a rupture will bring about, and again these are seen at various levels, from the scientific (Drexler and Smalley 2003; Johnson 2010), to the socioeconomic (Hunt, Mehta, and eds. 2008; Anderson et al. 2009), and to the regulatory levels (Bosso 2010).

2 General Philosophy of the Book

I have mentioned that I draw on institutional theory to analyze the pattern of interactions in which various stakeholders act collectively as institutional entrepreneurs and shape nanotechnology institutional emergence. It is now time to define what I mean by *nanotechnology institutional emergence*. One first step is to define institutions. Drawing from the work of Meyer and Rowan (1977), DiMaggio and Powell (1983), Tolbert and Zucker (1983), and Greenwood et al. (2008) and integrating their definitions, I offer my own interpretation of the notion of institutions: Institutions are ensembles of structured conventional social practices that have acquired a taken-for-granted ontological status that allows them to set conditions on action. Two very important aspects of institutions are that they are "permanent" and "universally acknowledged" social structures.

"Permanent" does not mean that institutions are everlasting—it would be impossible to have new institutions if that were the case, and the notion of institutional change would not mean anything. By "permanent," I consider that institutions have a relative persistence in time, a persistence that outlasts fads and trends. Bureaucracy has been and still is to a certain extent an institution. It corresponds to a norm, a standard of behavior, a set of practices that is widely understood (meaning that everyone understands what it corresponds to) and accepted. It was one of the most widespread organizational forms in the Western world of the twentieth century (Weber and Gerth 1958; Weber 1997). Yet organizational structures are shifting, and even if bureaucracy still appears as a set of organized practices fitting a well-defined social, legal, and economic framework, bureaucracy is no longer the dominant model that it used to be, and it has undergone change (Lounsbury and Carberry 2005; Beck 1992; Kallinikos 2003).

"Universally acknowledged" does not mean either that everyone on Earth places a given institution on the same value scale. By "universally acknowledged," I mean that institutions have a form of *cultural identity* that makes people recognize what they are whatever the various organized forms they take. In French, "armed forces" comes under one generic word: *L'armée,* which serves to make the symbolic meaning stronger than the somewhat hazy term of "armed forces." However, whatever the name, whatever people think of its value, when the French read or hear about "armed forces" (*l'armée*), they immediately identify what this refers to (with symbols such as uniforms, discipline, national identity, and so forth). "Armed forces" constitutes an ensemble of structured conventional social practices. It has a taken-for-granted ontological status that allows it to set conditions on action: apart from the few nations that were forbidden to keep armed forces at the end of the Second World War, it is a universally accepted fact that nations keep armed forces. "Armed forces" constitutes an institution whose cultural identity is probably one of the most widely shared on Earth. Yet the U.S. Army is quite different in its organizational from, say, the Ugandan armed forces. The family is also an institution, but the outlines of this institution are not the same everywhere. Family laws, structures, and organizations vary among different cultures. In Western societies, the family institution has undergone huge changes. Yet the cultural nature of this organized form of social links is universally known; it is an institution.

Can nanotechnologies delineate an institution? The straightforward answer to this question is no, and I will explain why. The social practices that delineate institutions are usually organized around a certain technology, or set of technologies, that has a specific purpose. This may be weapons in the case of the armed forces, reproduction in the case of the family, or the tools of organized rationalization (accountancy books, various types of records or reports, ranking tools, etc.) in the case of bureaucracy. Until 2003, there was a consensus to present nanotechnologies as a single technology. An archival search of 192 papers written between 2001 and 2003—which covers the majority of the papers written during that period after we eliminate duplicates—shows that the word *nanotechnology* was almost always used in the singular at that time.

The most widely accepted explanation for this is that the first stakeholders (essentially scientists and politicians) purposely created hype around "nanotechnology." The reasons for this will be debated at a later time. One of the means for creating hype was to use a word in the singular as a way to give existence to an otherwise completely immaterial concept. "Nanotechnology" started out as a visionary program, *defined by the effects it would produce.* The first official documents about it were released by the U.S., and they served to describe a vision of a unifying technology that would bring about a revolution. This revolution was presented as the latest in a succession of revolutions: the Industrial Revolution had lasted two hundred years; followed by the electronics revolution, which had a likely working life of seventy-five years; and then by the biotechnology revolution, which, although hardly yet on its feet, was already prophesied to be supplanted by (or perhaps to morph into) the nanotechnology revolution. A few quotations illustrate this mechanism:

> Nanotechnology's relevance is underlined by the importance of controlling matter at the nanoscale for health care, the environment, sustainability, and almost every industry. There is little doubt that the broader implications of this nanoscience and nanotechnology revolution for society at large will be profound. (Roco and Bainbridge 2001, 9)

> Those who participate in the "nano revolution" stand to become very wealthy. (Roco and Bainbridge 2001, 19)

> The promise of nanotechnology to remake our world seems virtually infinite. So the first thing to say is that if—as is variously claimed—nanotechnology is going to revolutionize manufacturing, health care, travel, energy supply, food supply, and warfare, then it is going, as well, to transform labor and the workplace, the medical system, the transportation and power infrastructure, the agricultural enterprise, and the military. (Roco and Bainbridge 2001, 57)

In 2001, the U.S. created a single program centralizing the major research efforts relating to "nanotechnology." This was the National Nanotechnology Initiative (NNI). Prior this program, nanotechnology R&D was a haphazard smattering of various unconnected initiatives, in the U.S. and everywhere else. The creation of the NNI was followed by the publication of the S 189 Act, the full title of which was *21st Century Nanotechnology Research and Development Act* (U.S. Congress 2003). Tellingly, both landmarks documents referred to nanotechnology in the singular.

In the wake of the U.S.'s action, other countries, as well as regional powers such as the EU, began to recognize the importance of coordinating nanotechnology research and establishing it as a field. The first such document produced by the EU was *Towards a European Strategy for Nanotechnology* in May 2004 (European Commission 2004). There again, the word *nanotechnology* took the singular form.

So it can be said that the birth of nanotechnology was first orchestrated by politicians and scientists in the U.S. as a way to create a nanotechnology institution. This institution envisioned that it would be a global institution and that it would revolutionize society for the better—by acting on the structure of matter. This was the way to create a new ensemble of structured conventional social practices that would set conditions on action.

But as more and more research was carried out and the field began to emerge, it became clear that *nanotechnology* as a singular word did not mean anything: the modification of matter at the atomic scale took so many different forms, rested on so many different approaches, and led to so many different outcomes that it could not be assimilated to one single technology. As Nordmann (2007) explained, the era of nanotechnologies is rather like the advent of plastic. Plastic was the first nonnatural material to be mass produced for the general consumer. Now it is everywhere around us, and even within some of us (as a result of certain surgical procedures). Plastic has diffused across most of the sectors of activity and become ubiquitous. With nanotechnologies, the situation appears to be very similar, with an even greater extent of diffusion. Based on a list of sectors directly affected by the development of nanotechnologies compiled by the European Commission DG Research—G4: Nano S&T and Converging Science and Technologies, here is my own illustration of the pervasiveness of nanotechnologies in 2007 (Figure 2.1).

Although plastic is all-pervasive, and although a great deal of our social practices have changed with its arrival on the market in various forms, one cannot say that plastic is an institution. It has *changed* our practices, but it does not correspond to an ensemble of practices. The same can be said of nanotechnologies. In spite of the initial thrust that presented them as a unified technology, they are now referred to as nanotechnologies in the plural. A year after the EU released its first landmark document entitled *Towards a European Strategy for Nanotechnology*, it produced a second report called *Nanosciences and Nanotechnologies: An Action Plan for Europe 2005–2009* (European Commission 2005). It was but one example of the sea change that was taking place. Today, few documents refer to nanotechnology in the singular, and just as plastic is not an institution, nanotechnology, or even nanotechnologies, does not correspond to an ensemble of social practices, and does not constitute an institution. We might go further and

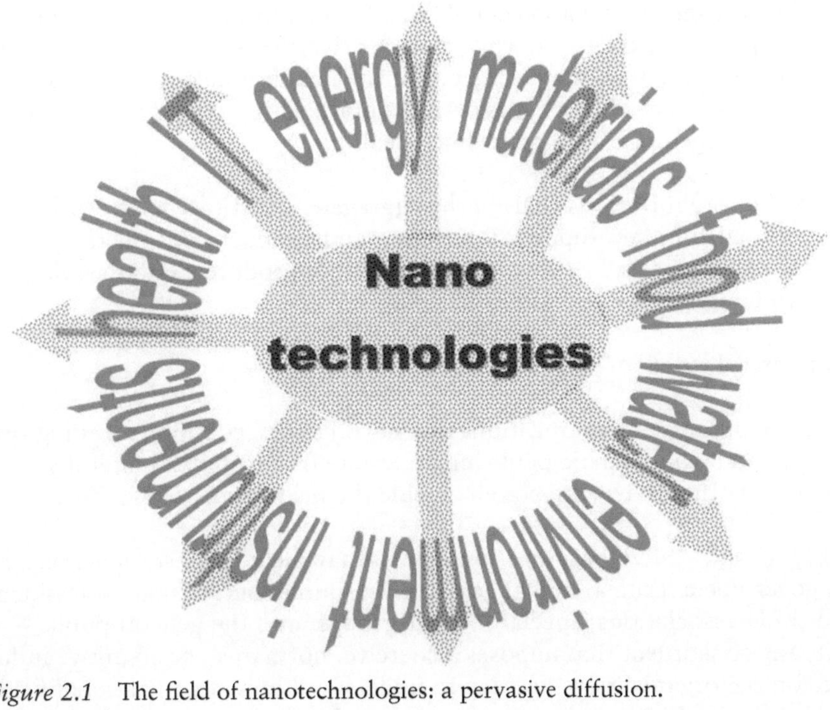

Figure 2.1 The field of nanotechnologies: a pervasive diffusion.
Source: author.

say that the only common thread in all *nano* technologies—and it is a major one—is that they deal with the most basic components of matter. But if we wanted to stretch it a bit, we might say that all living and inert substances of the universe are made of atoms anyway, so why consider that something special is happening? That is, is there really anything to write about?

I argue that there *is* something to write about, because something special *is* happening with nanotechnologies. Scientists are modifying the structure of matter, and in so doing, they are completely changing the human paradigm. *Nanotechnologies* are revolutionary, in the strongest sense of the term. They may not correspond to an institution, but they are shaping a new transversal field that is going to spread until it affects most human activities one way or another. This emergence is at the core of this study, because it brings about a whole new ensemble of structured social practices that I intend to study in detail.

INSTITUTIONAL EMERGENCE AROUND THE DEVELOPMENT OF NANOTECHNOLOGIES

I explained in the introduction that the aim of this book is to explore the development of nanotechnologies to show that the institutional changes that coexist with this development are a sign of deep societal transformation. I have just clarified how nanotechnologies do not constitute an institution. In the next pages, I will explain why I consider that they constitute a field. I will then demonstrate how the emergence of the nanotechnology field coincides with the emergence of new types of institutions. This is what I mean by institutional emergence. DiMaggio defined fields as

> sets of organizations that, in the aggregate, constitute an area of institutional life; key suppliers, resource and product consumers, regulatory agencies, and other organizations that produce similar services or products. (DiMaggio 1983, 148–49)

Scott later defined an organizational field as

> a community of organizations that partakes of a common meaning system and whose participants interact more frequently and fatefully with one another than with actors outside the field. (Scott 1995, 56)

For Hoffman (1999), organizational fields may include constituents such as the government, critical exchange partners, sources of funding, professional and trade associations, special interest groups, and the general public—in fact, any constituent that imposes a coercive, normative, or cognitive influence on the organization. Seeking to understand how organizational fields and institutions coevolve, he suggests that fields form around central issues, not around markets or technologies:

> The notion that the field forms around a central issue—such as the protection of the natural environment—rather than a central technology or market introduces the idea that fields become centers of debate, where competing interests negotiate over issue interpretation. As a result, competing institutions may lie within individual populations (or classes of constituencies) that inhabit that field, what I define as "situated" institutions. (Hoffman 1999, 353)

This is precisely what the rest of this book is about: showing that the field of nanotechnologies is shaping around a central issue—the realization of the importance of sustainable development. Nanotechnologies are indeed centers of debates where the tenants of competing interests negotiate over issue interpretation. So when I speak of the institutionalization of nanotechnologies—or nanotechnology institutional emergence—I refer in fact to the emergence of new types of institutions around the development of the nascent field of nanotechnologies. To clarify the complex set of interactions that are at play, I explore the roles of five groups of constituents whose actions impose competing "coercive, normative, and cognitive influences." These five groups of constituents are scientists, entrepreneurs, commentators, end users, and policy makers. The term *commentator* is meant to have a very generic meaning: it may be someone who observes and comments on something, but also someone who reports and analyzes events. Therefore it includes civil society representatives and NGOs, as well as journalists and other stakeholders reporting on the field. These five groups of constituents act as institutional entrepreneurs because they mobilize resources to modify existing institutions or to create new ones to suit their purposes around the development of nanotechnologies (Figure 2.2).

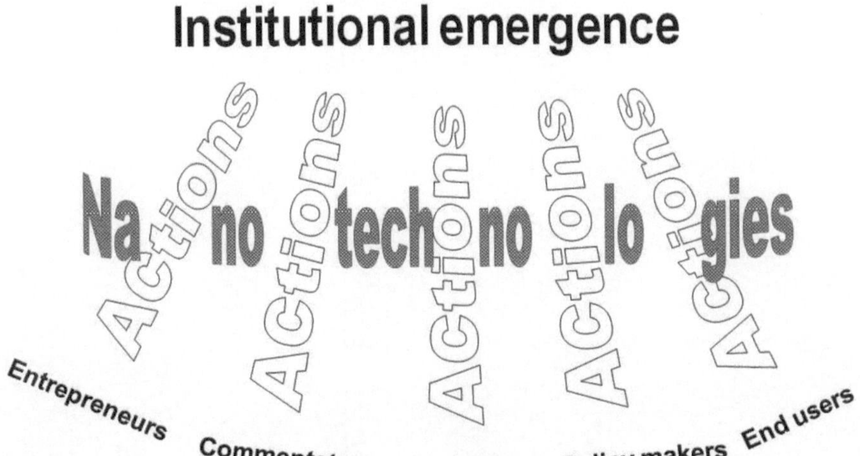

Figure 2.2 Nanotechnology institutional emergence.
Source: author.

Part II

Indicators of Nanotechnology Entrepreneurship

3 Context and Definitions

As a first step in the study of the emergence of the nanotechnology field, this chapter provides a snapshot of salient features of nanotechnology entrepreneurship. This will serve in later chapters to explore connections with institutional emergence and build the main argument of the research: that a key element to understanding the relation between the emergence of the nanotechnology field and the creation of new institutions or modification of existing ones is the awareness of the importance of the concept of sustainable development.

Since the term *entrepreneur* was used for the first time in an economic context in 1755, purportedly by Cantillon (1952), the concept of entrepreneurship has spawned a growing amount of literature, much of it trying to define what entrepreneurship really is. Schumpeter was the one to introduce in the 1930s the notion that entrepreneurship is the process of carrying out new combinations, which may take the form of new products, processes, markets, organizational forms, or sources of supply:

> Whatever the type, everyone is an entrepreneur only when he actually carries out new combinations and loses that character as soon as he has built up his business, when he settles down to running it as other people run their business. (Schumpeter 1934, 78)

Entrepreneurship has since been described as occurring when an individual or a group of individuals, in association with an existing organization, creates a new organization or instigates renewal or innovation within that organization with the view to generate new businesses (Collins and Moore 1970; Gartner 1988; Sharma and Chrisman 1999; von Hippel 1977). Building on the work of these scholars, and on Scott Shane's general theory of entrepreneurship (Shane 2003), I use the classic definition of the word—*the creation of new products and services out of the identification, evaluation, and exploitation of either new resources or the combination of existing ones, in order to make a profit*. Definitions of entrepreneurship vary, tend to apply to more and more forms of human activity, and do not always consider profit making as a discriminating feature. For example, in their exploration of social entrepreneurship (Dorado 2006; Wei-Skillern et al. 2007), some researchers disconnect profit making from the engine that drives the mobilization or recombination of resources. I deliberately place myself in an economic context of tangible value creation, and my understanding of entrepreneurship is based on the notion of profit seeking. This positioning may be debatable, but it allows me to outline a form of emergence of nanotechnologies that can be clearly defined, based as it is on quantifiable values. Moreover, nanotechnology is recognized as a promising new growth innovator for the decades to come (Knol 2004; Paull et al. 2003; Roco 2001; ten Wolde 1998), and the studies reaching these conclusions are all based on the creation of new profit-making markets.

Building on the aforementioned definition, I focus on the following form of entrepreneurship, which I call "nanotechnology entrepreneurship:" *when organized actors (such as university spinouts*[1]*, start ups, or even parts of established firms) use collaborative relationships with existing institutions (such as business angels, institutional funding agencies, or corporate funding) to generate new organizations creating future goods and services around the emerging technological field of nanotechnologies*. This definition serves as the framework for my study, and it is on this form of entrepreneurship—which includes *intrapreneurship* as defined by Pinchot (1985) and Pinchot and Pellman (1999)—that I build my exploration of institutional emergence.

Just as there is a large body of literature attempting to define exactly what entrepreneurship is, the attempts to measure entrepreneurship have led to many projects and theoretical models (Blanchflower and Oswald 1998; OECD 1998; OECD and Davis 2007; Bosma et al. 2010). Among these, some models account for entrepreneurship by combining R&D expenditure and patent numbers with other variables (Aghion and Howitt 1997; Darby, Liu, and Zucker 2004; Shane 2003). I choose to build on this theoretical approach for a couple of reasons: nanotechnology is an emerging field that is based on technological innovation whose growth relies heavily on R&D and patents. Also, nanotechnology entrepreneurship connects closely with university research, as is apparent with the rise of university spinouts and partnerships between business and universities (The Institute of Nanotechnology 2004). Therefore, I base my outline of entrepreneurship as defined above on three indicators: an analysis of nanotechnology R&D funding; a mapping of nanotechnology research centers and infrastructure; and, finally, a study of nanorelated patent activity. I conclude by looking at how these three indicators relate to nanoproducts reaching the market.

4 R&D Architecture

Nanotechnology R&D funding started being documented on a systematic basis in the early 1990s. Before that time, although there were programs focusing on atomic manipulation, nanotechnology R&D as such was too small to be significant. In a period of about ten years, the total global investment in nanotechnology R&D, public and private, reached around 15 billion euros in 2006 (Sources: Cientifica, Lux Research Inc., European Commission, and NNI). This amounted to nearly 0.05 percent of the world gross domestic product (GDP), some €32.6 trillion in 2006, based on the gross world product (GWP) combined data of the CIA World Factbook, IMF, and World Bank. Different sources such as the National Science Foundation (NSF) and the National Nanotechnology Initiative (NNI) predict that the market for nanotechnology-based products will be worth at least a trillion euros in 2015, and the consultancy Lux Research Inc. reports that by 2014, $2.6 trillion in global manufactured goods will incorporate nanotech (Lux Research Inc. 2007). Based on a projected annual growth rate of the GWP of about 5 percent in 2006–7 (World Bank and National Bureau of Economic Research figures), the nanotechnology industry—which was still negligible in terms of marketed products in 2000—would then contribute about 3.5 percent of the GWP.

There has been much hype around the projected trillion-euro nanotechnology industry, and analysts rate nanotechnology as a very promising sector (MRG Multimedia Research Group Inc and Fuji-Keizei 2004; Paull et al. 2003; Gillis and Krim 2004). Indeed, the simple calculation in the preceding paragraph shows how promising the sector is. However, it also signals something else: the technology is likely to be very disruptive. A sector that—from hardly any marketed outcomes in 2000—is projected to affect 3.5 percent of the GWP fifteen years later inevitably brings about major changes on its way. The short time period during which this change is due to come about is another new element. Drexler (1991) and other scientists have noted the acceleration of the eras that have marked the evolution of mankind, from the agricultural era lasting several thousand years to the industrial era lasting a few hundred years, to what has been called the new Industrial Revolution—the information age—which has to this date lasted a few decades, and is giving now way to the nanotechnology era.

Global R&D spending in percentage of GDP - 2005	
Americas	2,3
Asia	1,8
Europe	1,8
World	2
Source: R&D magazine, Batelle, OECD, World Bank	

Chart A

Percentage of the total global R&D in 2005 and 2007	2005	2007
Americas	37,7	36,8
Asia	34,9	36,5
Europe	24,1	23,2
Other	3,3	3,5
World	100	100
Source: R&D magazine, Batelle, OECD, World Bank		

Chart B

Figure 4.1 Global R&D public spending in percentage of GDP. Source: data from CIA World Factbook, European Commission, OECD, International Monetary Fund (IMF), and World Bank. Author's compilation and chart.

It is no easy task to obtain accurate data concerning nanotechnology R&D funding, particularly private funding. When assessing R&D funding in general, analysts usually consider that the total amount of public funding is matched by private funding. In nanotechnologies, as we will see, there are some important differences between public and private funding of nanotechnology R&D, depending on the countries. Public funding of nanotechnology R&D has been monitored on a fairly systematic basis since the early 2000s. Yet the comparison of the various available data sources

Public nano R&D expenditure in percentage of GDP – 2006	
Americas	0,013
EU	0,013
Asia	0,018

Figure 4.2 Global nanotechnology R&D public spending in percentage of GDP.
Source: author.

reveals important differences, indicating that figures are to be taken more as indications of tendencies than at their face value. With this word of caution in mind, here is a panorama of nanotechnology R&D public funding in percentage of country GDP (gross domestic product). The data used in the following analysis comes from five major sources: CIA World Factbook, European Commission, OECD, International Monetary Fund (IMF), and World Bank. We begin with Figure 4.1, which provides an account of public R&D funding in general.

Figure 4.1 shows that *on the whole,* countries spend some 2 percent of their GDP on R&D in general. It also shows that globally, Europe lags somewhat behind Asia and the Americas in terms of its share of global R&D funding. In 2007, it contributed less than 25 percent of the overall R&D funding, whereas Asia contributed over a third and the Americas even more, with nearly 40 percent of the world effort. Next, Figure 4.2 provides an outline of public *nano* R&D expenditure. The chart was obtained by compiling data from the CIA World Factbook and World Bank for countries' GDP, and from Lux Research Inc. (2007) for nanotech public funding.

Figure 4.2 shows that nano R&D corresponds in fact to a very small fraction of R&D in general. As a reminder, in 2006 the total of nanotechnology R&D funding—public and private—amounted to some 0.05 percent of the world GDP, a very small part of the world's assets thus far. One first observation is that there is a discrepancy between the hype surrounding the potential of nanotechnologies and the actual sums that have been invested in them. However, these general statistics mask the fact that different countries do not give the same relative value to nanotechnologies. Unsurprisingly, the countries investing most in nanotechnology public R&D are highly developed countries in terms of new technologies. This may be an explanation why two small countries, Ireland (ranked 120th in the world in terms of geographical area) and Israel (ranked 152nd), have decided to spend, in relative terms, a much greater portion of their public resources on funding nano R&D than other nations. Ireland's GDP is 32nd in the world and Israel's, 45th in the world. However, Israel comes first in terms of national R&D spending (general R&D), with a budget reaching 4.6 percent of GDP in 2004, and it also comes out on top of the list in terms of national public nanotechnology R&D funding as a share of GDP (0.1 percent in 2006).[1] The Israel National Nanotechnology Initiative (INNI) was established in 2001. It explained that the significant amount of nanotechnology R&D funding was due to the fact that Israel was recognized the world over as a leading source of knowledge and innovation in advanced research and technology, particularly in the fields of communications, electronics, computer software, networking, defense, security, and life sciences industries. In addition, the initiative stated that Israel's workforce was highly educated, with over 20 percent of Israel's working population holding university degrees. For each 10,000 employees, 140 were scientists or technicians, and 135 were engineers, a much better ratio than in the U.S., Japan, or Europe. Israel's academic, business, and government leaders recognized nanotech as a key platform for ensuring Israel's continued R&D excellence for decades to come, and the INNI considered that nanotechnology was a natural enabler for Israel, with the small size of the country an advantage because it meant sharper focus, more efficient use of funds, fewer commercial obstacles, rapid prototyping and testing, and higher quality standards. An article from the *International Herald Tribune* offered another explanation for Israel's substantial participation: the role of Shimon Peres, the incumbent president of Israel since 2007. Shimon Peres was born in 1923 and had participated in the development of Israel's nuclear option in the 1950s. He then wanted Israel to invest in nanotechnology because he was convinced that it was key to Israel's economic future. "Nuclear power is destructive but nanotechnology is all about constructing the world anew," Peres said in a telephone interview to a *Tribune* journalist. "It gives us the opportunity to change everything: to make computers the size of a head of a pin, material hundreds of times stronger than steel, robots the size of a single hair on your head that can travel inside your body and attack cells like those that cause cancer" (Schenker 2003).

In spite of historical and geographical differences, Ireland's position regarding nanotechnologies was rather similar to that of Israel's. Its focus on nanotechnology R&D resulted from two major factors. First, it had a strong university research network in high tech. This network spawned indigenous start-ups that emerged from university-based research groups, and these attracted international venture capital as early as the mid-1990s.

Nanotechnology quickly became a key enabling technology in industrial settings that were very promising: electronics, pharmaceuticals, energy, chemicals, and consumer products. In addition, Ireland was in 2004 one of the most competitive economies in the world, with a GDP of €34,000 per capita, just behind the U.S. Its attractive tax incentive of 12.5 percent for corporate tax, combined with a well-educated, skilled workforce, attracted strong foreign direct investment (FDI), and a number of MNCs (multinational corporations) set up their European headquarters in Ireland. Out of 1,200 companies based there, 1,000 were in the IT (information technology) sector. And, nine of the top ten pharmaceutical companies and thirteen of the top twenty medical devices companies had bases in Ireland. This combination created an environment that was particularly adapted to the pursuit of high-tech ventures, and nanotechnology was identified as the key area to support.[2]

It is generally admitted that, *in general,* private funding matches public funding. However not every nation adopts the same approach. In the U.S., public funding comes essentially from two sources, federal funding and funding from the states. In the EU, public funding also comes mainly from two sources, this time from the European Commission and from the individual nation-states. Public funding tends to be more significant in the EU than in the rest of the world. Over the period 2004–6, it amounted to 66.6

percent of nanotechnology R&D funding in the EU, whereas it was 45.6 percent in the U.S. and 36.8 percent in Japan (Figure 4.3).

Figure 4.3 confirms that while the U.S. has a fair balance between private and public funding, private investment is lagging behind public funding in Europe. This trend is not specific to the development of nanotechnologies but is part of a general pattern whereby Europe has a tradition of greater public funding for research in new technologies (Hullmann and European Commission 2006). According to the Hullmann and European Commission (2006) report, whether public funding brings about better quality of research than private funding is not clear, but private-funding research tends to have more short-term objectives of profitability than public research. The report states that the high level of public funding of nanotechnology research is likely to have a positive impact on European excellence in science and technology, but has no direct correlation with successful technological implementation and the translation into commercially successful products, as these projects depend on the integration of industry. A greater share of *private* funding in R&D means greater industrial integration, and this combination is more conducive to entrepreneurship. The European Commission also notes in the same report that the importance of public funding in relation to private funding leads to an additional conclusion: that research into the societal implications of nanotechnologies is valued more in Europe than in the U.S.

Public-research funding strategies may have important repercussions on the economic development of nanotechnologies, as it is felt that nanotechnologies will be accepted by the general public if they are seen as providers of

Global nanotechnology R&D funding, in million euros, 2004-2006

	Private	Public	percentage public money
JAPAN	3640	2120	36,8
US	4500	3770	45,6
RoW	1200	1830	60,4
EU	2350	4680	66,6

Figure 4.3 Proportion of public funding of nanotechnology R&D.
Source: author.

new solutions, not of new problems. One great advantage of publicly funded research is that it can steer research politically—notably through the definition of priority areas—to stay in line with public expectations. By influencing the direction of nanotechnology research toward areas such as the safety aspects of nanotechnology, new environmental solutions, or new medical devices, it can correspond better to societal expectations. Whether doing so translates into a positive economic impact is not entirely clear, but potential backlashes coming from public rejection of a new technology are more likely to be avoided.

Another important feature of public funding for nanotechnology is that the European Commission, as a body, accounts for about one-third of the general EU budget of nanotechnology public R&D funding, the other two-thirds coming from individual nation-states. In the U.S, the reverse is true: federal agencies account for nearly 70 percent of all public nanotechnology R&D funding, while a small one-third comes from the states. This inverted positioning confirms the fact that the EU is still in a phase of integration whereby nation-states have more power than the global entity, whereas the U.S. acts as an integrated body with an overpowering federal power.

Looking now specifically at *public* R&D funding, we see an important difference between the U.S. and the EU: it is the way the NNI and the European Framework Programmes are financed. I will start with the NNI to explain the difference. The NNI is an interagency program that coordinates federal nanoscale research and development activities and related efforts among various participating entities. It was created in 2001 by the U.S. government, and its activities were defined in the 21st Century Nanotechnology Research and Development Act of 2003 (U.S. Congress 2003). Six agencies developed the original 2001 proposal. They were the Department of Defense (DOD), the Department of Energy (DOE), the National Aeronautics and Space Administration (NASA), the National Institutes of Health (NIH), the National Institute of Standards and Technology (NIST), and the National Science Foundation (NSF). In 2010, the NNI gathered a total of 25 agencies. Thirteen contributed to its budget, but their participation was not evenly shared. Of a total budget of approximately $1.7 billion in 2009, the Department of Defense was the major contributor with 26.9 percent of the overall budget, followed by the National Science Foundation with 24 percent, the National Institutes of Health with 20 percent, and the Department of Energy with 19.5 percent. Together, these four agencies accounted for over 90 percent of the NNI budget (Figure 4.4).

The agencies fund the areas of research that are the most likely to strengthen their position altogether. This is made clear by Figure 4.5, which shows how each of the NNI participating agencies splits its funding along the major Program Component Areas (PCAs) it identified. The differences between the ways each agency finances the various PCAs are quite significant.

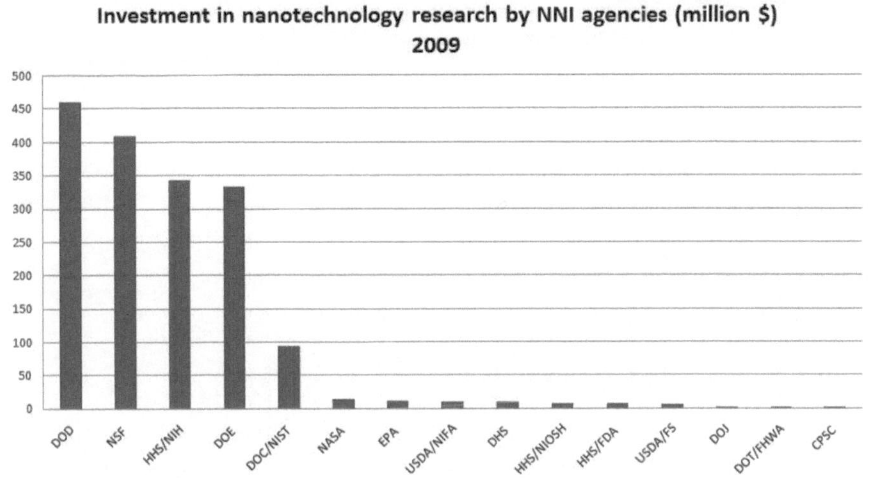

Figure 4.4 NNI agencies funding nanotechnology research and their compared contributions.
Source: NNI 2010. Author's compilation and chart.
Acronyms:
DOD—Department of Defense
NSF—National Science Foundation
HHS/NIH—Health and Human Services/National Institutes of Health
DOE—Department of Energy
DOC/NIST—Department of Commerce/National Institute of Standards and Technology
NASA—National Aeronautics and Space Administration
EPA—Environmental Protection Agency
USDA/NIFA—United States Department of Agriculture/National Institute of Food and Agriculture
DHS—Department of Homeland Security
HHS/NIOSH—Health and Human Services/ National Institute for Occupational Safety and Health
HHS/FDA—Health and Human Services/Food and Drug Administration
USDA/FS—United States Department of Agriculture/Forest Service (Department of Agriculture)
DOJ—Department of Justice
DOT/FHWA—Department of Transportation/Federal Highway Administration
CPSC—Consumer Product Safety Commission

With the EU, the system is quite different. The EU is an association of independent member states (twenty-seven in 2010). This association finances common policies throughout the union with an agreed budget that is financed directly by member states and that has been set at about 1 percent of the EU's GDP (€116 billion for year 2007). Within this budget, the funding of research amounts to 5 percent of the total. Due to the variance in relative economic health, countries do not all contribute the same

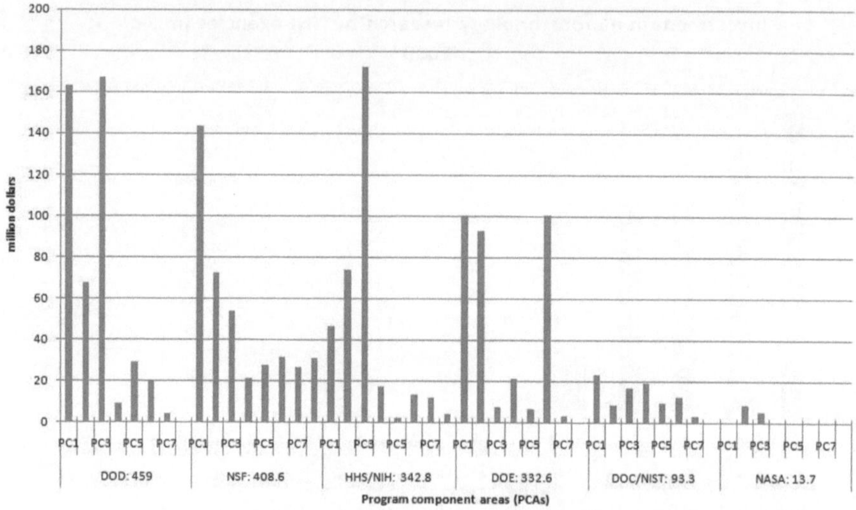

Figure 4.5 NNI investments by agency and by Program Component Areas in 2009. *Source*: author; compilation based on NNI data. *Note*: the figure appearing after each agency is its total investment in million USD (United States dollars).

Key to Program Component Areas:
PCA #1: Fundamental Phenomena and Processes
PCA #2: Nanomaterials
PCA #3: Nanoscale Devices and Systems
PCA #4: Instrument Research, Metrology, and Standards
PCA #5: Nanomanufacturing
PCA #6: Major Research Facilities and Instrumentation Acquisition
PCA #7: Environment, Health, and Safety
PCA #8: Education and Societal Dimensions

percenage of their GDP to the EU, and it is the pool of all their contributions that funds public research centers and infrastructures.

Figure 4.6 shows the contributions of the four major EU contributors (Germany, France, Italy, and the United-Kingdom) and of two others that are used for comparison (Belgium and Ireland).

These four major EU contributors are called net contributors because they contribute more to the EU budget than they receive in various financing. Conversely, Belgium and Ireland are two countries that receive more from the EU budget than what they contribute. Public research is not structured along the national lines of participating countries in relation to their involvement in the EU budget. Ireland and Belgium are among the six most important EU nanoresearch centers and structures, yet they are not at all among the top contributors of EU funding.

The Framework Programmes for Research and Technological Development, also called Framework Programmes (FPs), are funding programs

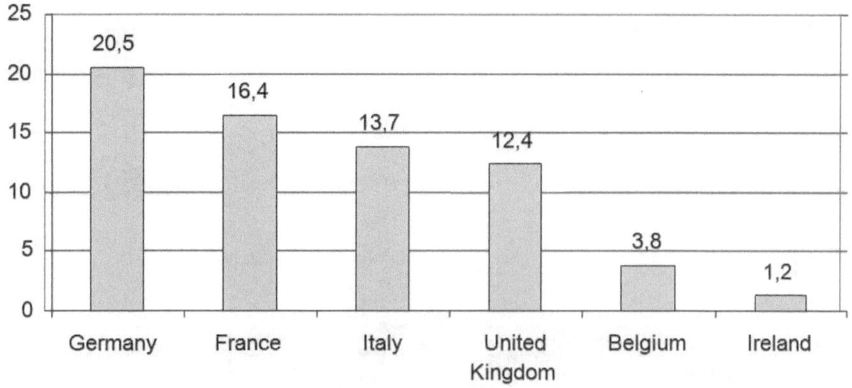

Figure 4.6 Contribution of a few EU countries as a percentage of total EU budget—2006.
Source: European Parliament, 2006 budget. Author's compilation.

created by the European Union in order to support and encourage research at the European level. The detailed objectives and actions vary from one funding period to another. Nanotechnology research started appearing in Framework Programmes with FP4, in 1996. Since then, the European Commission released three more Framework Programmes, FP5, 6, and 7, each of them lasting between three and four years. FP7, the program with the longest life and the largest budget so far, is running from 2007 to 2013. As appears on Figure 4.7, nanotechnology funding really started taking off in 2004.

The total budget of FP6 (from 2003/4 to 2006) for nanoscience and nanotechnologies (N&N) was nearly €1.4 billion, shared over 550 projects. For the same time period (2004–6), the budget of the U.S. National Nanotechnology Initiative was $3.5 billion dollars, i.e., €2.7 billion (€1~$1.30) (Source: NNI). These similar budgets covered different approaches. While the European Commission funds about a third of EU public nanotechnology research, with the rest coming from the funding of individual states, the NNI accounts for nearly 70 percent of all U.S. public nanotechnology research funding. So, even if both the NNI and the Framework Programmes were organized along seven major program areas in 2006, the foci of the program areas were different (Figures 4.5 and 4.8).

Eleven percent of the NNI budget went to infrastructures and equipment acquisition, to which could be added 5 percent going to research into metrology and standards. In monetary terms, the U.S. spent some €200 million euros on these aspects of the development of nanotechnologies, while the EU FPs did not mention infrastructures in their program areas. So the U.S. chose to focus on setting the framework for nanotechnologies, not only by building vital infrastructure, but also by giving itself the means to set up the international standards that would then be used globally. As

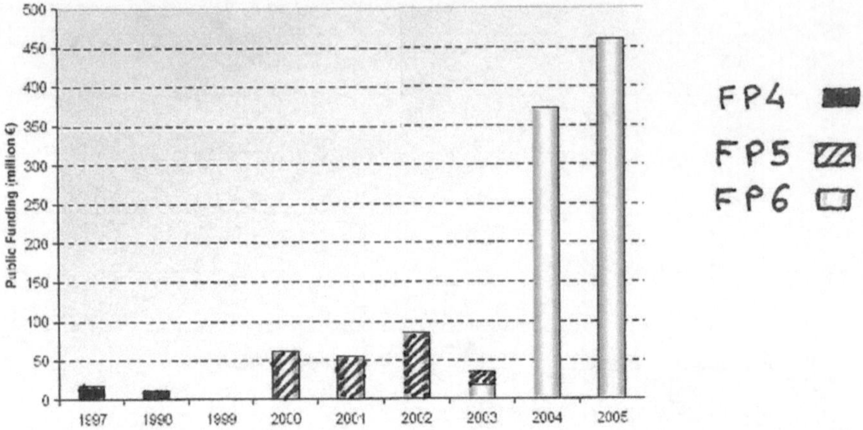

Figure 4.7 Evolution of EU Framework Programmes (FPs) funding devoted to nanotechnology R&D.
Source: European Commission, adapted.

Garud, Jain, and Kumaraswamy (2002) explained at length, infrastructure and standardization are two very important steps in the institutionalization of new technologies.

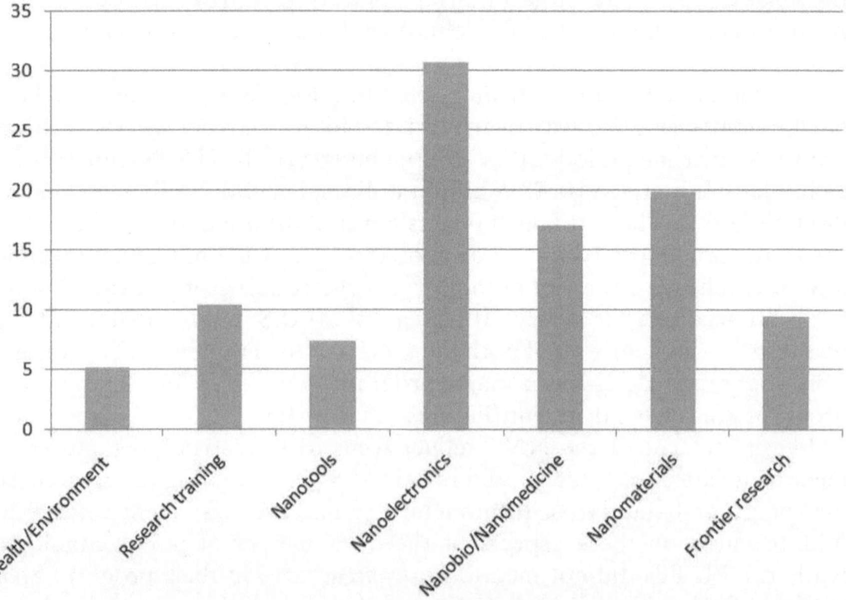

Figure 4.8 FP6 nanotechnology R&D funding by area in percentage.
Source: European Commission. Author's compilation and chart.

At the same time, the EU still devoted about 10 percent of its budget to research training. Even if this share of the budget was gradually decreasing, demonstrating that the goal was nearly attained, several EU reports still insisted on the necessity to harmonize the standard of nanotechnology research in the EU. Indeed, EU enlargement—accession of new member states—goes hand in hand with the objective of bringing all the EU countries toward a similar technological advancement. The European Commission calculated that there were, on average, 30 percent fewer researchers for every active one thousand persons in the EU than in the U.S. (European Commission 2003). This meant that some countries had a lot of catching up to do, which explains the need to expend adequate amounts to achieve European technological integration. The fact that the NNI did not mention research training seems to indicate that the U.S. was already past the stage of technological unification.

Over the period 2004–10, the NNI investment by Program Component Areas showed great consistency over time, while FP5, 6, and 7 reflected a constant refocusing. This is another illustration of the fact that, technologically speaking, the U.S. is much better integrated than the EU, where the construction and integration processes are still ongoing. One example of a refocusing of the EU Program Component Areas was the appearance in 2000 (FP5) of a new area called Environment and Health. It grew fast, and more than 5 percent of the FP6 budget was devoted to it. It confirmed the European strategy as expressed in its founding document to

> address any potential public health, safety, environmental and consumer risks upfront by generating the data needed for risk assessment, integrating risk assessment into every step of the life cycle of nanotechnology-based products, and adapting existing methodologies and, as necessary, developing novel ones. (European Commission 2004, 3)

In 2008, the European Commission produced a compilation of past and ongoing research projects associated with FP6 or FP7 in the area of possible impacts in health, environment, and safety of nanoparticles. The compilation, entitled *EU Nanotechnology R&D in the Field of Health and Environmental Impact of Nanoparticles*, was a first of its kind and contained information on 106 projects (European Commission 2008).

5 Research Centers and Infrastructure

Academic studies have shown the importance of research as a trigger of entrepreneurship in high-technology ventures (Wright et al. 2007; Zucker and Darby 1998). The following section maps nanotechnology research infrastructure based on various data sources coming mainly from the U.S. and the EU.

A study published by the EU nanotechnology gateway Nanoforum identified a total of 143 networks that offered support for collaboration and information exchange between members across 23 EU and associated states. Centers or networks were found in all EU and associated states apart from Croatia, Cyprus, Iceland, Liechtenstein, Malta, and Slovakia. Seventy-nine of these were national networks and sixty-four were involved in international cooperation. The report used the following eight broad categories to assess infrastructures and networks: all technologies; nanomaterials; electronics and systems; fundamental research; nanobiotechnology; analytical and diagnostics; engineering and fabrication; and energy. Most networks involved several disciplines, and over one-third of national networks covered them all. The most common type of thematic network focused on nanomaterials (28 percent). This was followed by electronics and systems (20.3 percent); fundamental research—mainly physics and chemistry—(16.8 percent); nanobiotechnology (15.4 percent); engineering and fabrication (4.9 percent); analytical and diagnostic tools (3.5 percent); and energy (2.8 percent). International networks tended to be more specialized (36.7 percent of national networks supported all disciplines, compared with 12.5 percent of international networks). However, in two areas (analytical and diagnostics, and energy), most of the relevant networks were national. Of the national networks, most (twenty-two) were coordinated from Germany, with nine from the UK, and four each from France, the Netherlands, and Poland (see Figure 5.1).

To map U.S. nanotechnology research centers and infrastructures, I compiled data from the NNI and various U.S. organizations such as the DOE or the DOD, as well as from the Nanobank, a large-scale multiyear project funded by the NSF to build a public digital integrated database of nanotechnology (Figure 5.2).

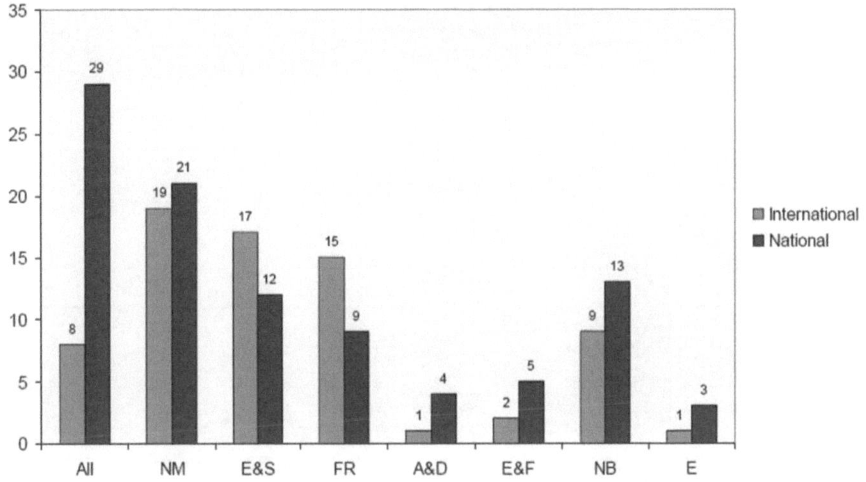

Figure 5.1 Breakdown of the 143 EU nanoscience and nanotechnology networks based on areas covered (showing international and national networks).
Source: Nanoforum, European Nanotechnology Infrastructure and Networks, 2005; adapted. Key: NM nanomaterials, E&S electronics and systems, FR fundamental research, A&D analytical and diagnostics, E&F engineering and fabrication, NB nanobiotechnology, E energy.

The research centers and networks are mostly located on the West Coast and eastern part of the U.S., with very few in the center and midwestern states.

The comparison highlights a vision of research that differs greatly between the EU and the U.S. As was mentioned previously, in the EU about one-third of nanotechnology R&D public funding comes from the European Commission's budget, which is itself a pool of contributions emanating from all the member states. The other two-thirds come from national public funds. In the U.S., nearly 70 percent of all public nanotechnology R&D funding comes from the NNI, which is a federal interagency program funded mainly by six agencies. So in the EU, public research tends to be structured along the lines of nations, while in the U.S. it tends to be structured along the lines of agencies.

This different structural approach has several effects. As public funding in the U.S. is mainly backed by agencies, it is more likely to be focused on problem solving than in the EU, where it is mainly backed by national funds. Apart from the importance of national priorities or specificities, the strong individual clout of nation-states also means that integration of research agendas is far from being achieved in the EU. Another effect is that while public R&D funding of nanotechnology in the EU has no direct connection with particular departmental domains, that is not the case in the U.S., with the overwhelming participation of government organizations such as the DOD,

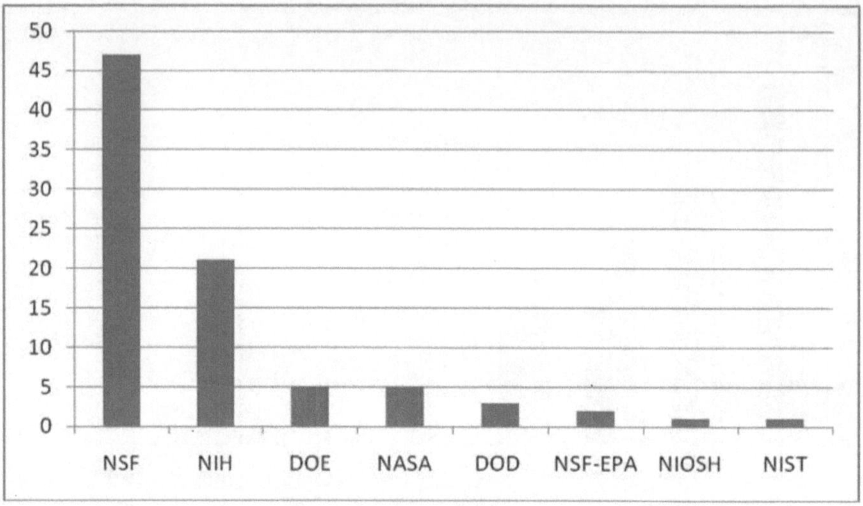

Figure 5.2 Number of nanotechnology research centers and infrastructures in the U.S. in 2009 (total: 85).
Source: Data from the NNI, DOE, and DOD; compilation and chart from author.
Acronyms:
DOD—Department of Defense
DOE—Department of Energy
NASA—National Aeronautics and Space Administration
NIH—National Institutes of Health
NIOSH—National Institute for Occupational Safety and Health
NIST—National Institute of Standards and Technology
NSF—National Science Foundation
NSF/EPA—National Science Foundation/
Environmental Protection Agency

the DOE, the NIH (which coordinate medical research), and the NSF (which supports research in science and engineering). Is this to mean that the U.S. funding system is more conducive to entrepreneurship than the European one? It would be difficult to support such claims given the complexity of the interactions that surround the development of nanotechnologies. However, these structural differences should not remain unnoticed.

6 Nanorelated Patent Activities

Several researchers link patent analysis with entrepreneurship, particularly in new technologies (Darby, Zucker, and Wang 2004; Akrich, Callon, and Latour 2006; Bonaccorsi and Grid 2005; Lemley 2005). According to Hullmann and European Commission (2006), patents reflect the ability of transferring scientific results into technological applications. Patents are also a prerequisite for the economic exploitation of research results (Chen and Roco 2008). In the following section, I analyze patent activities in the field of nanotechnologies, starting with a general outline, and then scrutinizing data coming from several patenting offices more finely. For this research I have worked mainly with the Nanobank and the Delphion databases. Delphion is part of the Thomson Corporation and provides access to intellectual property networks for the searching and analysis of patent data collections worldwide (see www.delphion.com). The lack of uniform definitions for *nanotechnology* means that it is difficult to identify the number of nanotech-related patents. In 2004, the USPTO (United States Patent and Trademark Office) created a new classification for nanotechnology patents—Class 977—precisely to address this problem. Class 977 defined nanotechnology patents narrowly by including only patents

1) whose subject matter is on the scale of approximately 1–100 nanometers in at least one dimension; and
2) that involve materials, structures, devices, or systems that have novel properties and functions because of their nanometric size.

However, in 2010, Class 977 included only 6,000 patents issued by the USPTO for the period 1980–2010, so Class 977 does not give a comprehensive picture of nanopatenting in the U.S., let alone in the world, which is why the Delphion database is preferable. This is currently the best available data bank of patenting activity for the regions of the EU and the U.S., although it is still far from being exhaustive. For example, Delphion lists under 'U.S. granted patents' *all* the patents issued by the U.S. Patent and Trademark Office, but under 'EU granted patents,' only the *European* patents issued by the European Patent Office. The database I use for this research comprises

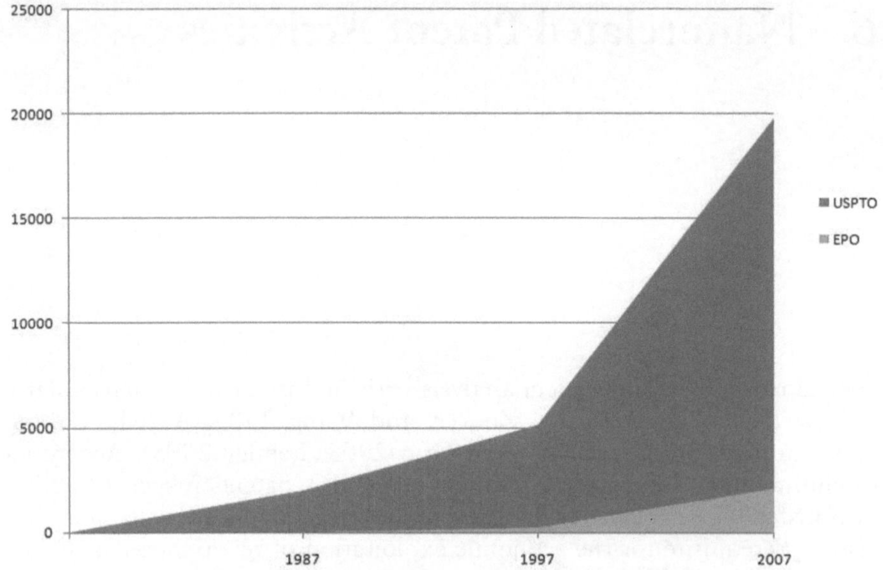

Figure 6.1 Evolution of the number of nanopatents in USPTO and EPO.
Source: Delphion; author's compilation.

patents from the USPTO (United States Patent and Tradesmark Office), EPO (European Patenting Organization), and WIPO PCT (World Intellectual Property Organization—Patent Cooperation Treaty). To search through patents, I use the generic search term 'nano.' Some critics argue that this can lead to exaggerated counts, particularly since it includes all nanocompounds, such as nanoseconds, for instance, which do not relate to space dimension. Even so, it is still likely the most efficient search term. I use the search term not only in the patent title but also within the complete text of patents.

As an introduction to my panorama of patenting activities in nanorelated fields, Figure 6.1 provides an outline of the evolution of the relative importance of the USPTO and the EPO as patenting organizations.

Figure 6.1 shows that the USPTO was still by far the agency of reference when it came to registering patents in 2007–8. Indeed, in July 2008, the number of patents falling under the heading 'EU granted patents' was just above a tenth of the number of patents under the heading 'US granted patents' (2,394 vs 22,028 patents). In their indicator-based study of the economic development of nanotechnology, Hullmann and European Commission (2006) looked at the evolution of nanopatenting in terms of the regions from which the patents originated. In 2003, they found that Europe was not only lagging behind the Americas (mainly the U.S. and Canada), but also behind Asia (mainly Japan and South Korea). Their analysis was based on the country from which the application for a patent was made and not on the nationality of the applicants. However, a trend toward more balance between the three regions became evident as the number of patents granted to American applicants began declining in 2002 while the other two regions were catching up.

In its study of the nanotechnology patent landscape from 1985 to 2005, Lux Research Inc. (2007) explained that while the filing for nanopatents kept rising at a brisk pace, the granting slowed down suddenly in 2005 because of an insufficient number of patenting officers at the USPTO. The number of patents granted increased by 26 percent in 2003 and by 22 percent in 2004, but only by 4 percent in 2005. The report attributed the slowdown to both the large number of applications and to the fact that new quality initiatives required more time for the USPTO to handle each application. According to the study, U.S. entities held an overwhelming 70 percent majority of U.S.-issued nanotech patents. Japanese entities were next at 9 percent, followed by German entities at 6 percent, and French and South Korean organizations with 3 percent each. The remaining 9 percent of nanotech patents were distributed among entities from thirty additional countries.

However, these findings were qualified by another report by Lux Research Inc. and Nordan (2005), which explained that in many specific promising application domains, researchers in other countries had begun to outpace the U.S. For instance, of the 70 patents for carbon nanotube display applications (potentially useful for large flat-panel monitors that could outperform LCD and plasma) issued by the USPTO through February 2005, only 17 percent were issued to entities based in the U.S., compared with 29 percent in Japan and 31 percent in South Korea. Another study on intellectual property in the nanotechnology economy coming from the UK-based Institute of Nanotechnology (Singh and Morrison 2007) revealed that in 2005–6, Japan was the most active country in nanoelectronics patenting. The main companies involved in the Far East were Fujitsu, with 62 nanoelectronic patent families, and Samsung, with 56 patent families. Other organizations active in this sector were Hewlett-Packard, Hitachi, Japan Science and Technology Agency, Infineon Technologies, and Philips.

These studies confirm that the core of nanotechnology patenting happens within the Americas, Southeast Asia, and Europe, leaving the other regions of the world as outsiders. There is fierce competition and much reorganization going on as patenting offices strive toward more harmonization of data. And, Delphion now also provides data from JAPIO (Japan Patent Information Organization). Asia is rising very fast, while Europe is lagging somewhat behind the other two regions in terms of patent activity.

For a more detailed outline of nanotechnology patenting, I combined data from the USPTO, the EPO, and the WIPO PCT. The WIPO PCT covers patent publications from over 175 countries, including the U.S. A Delphion search of the combined databases of the USPTO, the EPO, and the WIPO showed that there was in July 2008 a total of 7,201,129 granted patents and that out of these, 45,510 related to nanotechnologies.

Rank	Company	Headquarters	Nber of patents
1	IBM	USA	262
2	CANON	USA	165
3	SAMSUNG	SOUTH KOREA	137
4	HEWLETT PACKARD	USA	112
5	HITACHI	JAPAN	80
6	SEIKO	JAPAN	78
7	OLYMPUS	JAPAN	71
8	NANTERO	USA	68
9	INTEL	USA	67
10	ADVANCED MICRO DEVICES	USA	54
11	SONY	JAPAN	47
12	MICRON	USA	45
13	TOSHIBA	JAPAN	44
14	MATSUSHITA	JAPAN	37
15	3M	USA	37
16	D-WAVE-SYSTEMS	CANADA	35
17	KODAK	USA	35
18	NANOSYS	USA	35
19	MITSUBISHI	JAPAN	34
20	DUPONT	USA	33

Figure 6.2 Top twenty firms per number of patents granted by USPTO in 2010 under Class 977.
Source: USPTO; author's compilation and chart.

Nanotech patent ownership was extremely concentrated. Fewer than 100 entities owned half of all nanotech patents. Although there was a regular average of 1.5 to 2 patents per assignee, about a third of all assignees obtained between 3 and 12 patents, while the others had only 1. In 2010, of the 6000 patents granted by the USPTO under Class 977, approximately 4,000 went to companies. Within companies, the top ten accounted for a quarter of all nanotechnology patents awarded to the private sector. So nanotechnology research was also concentrated among a few firms, usually MNCs (Figure 6.2).

For a more comprehensive analysis of who patents most, I looked at the evolution of patent assignees over the last thirty years under five categories: business firms, universities, government agencies, individual claims, and others. The 'others' category gathers foundations that are not universities or research institutions per se, as well as patents that have no assignees (four patents came under 'unknown' in the column 'assignees' in 2007) (see Figure 6.3).

Figure 6.3 shows that fewer and fewer patents are granted to individuals. The number of patents being granted to government agencies remains more

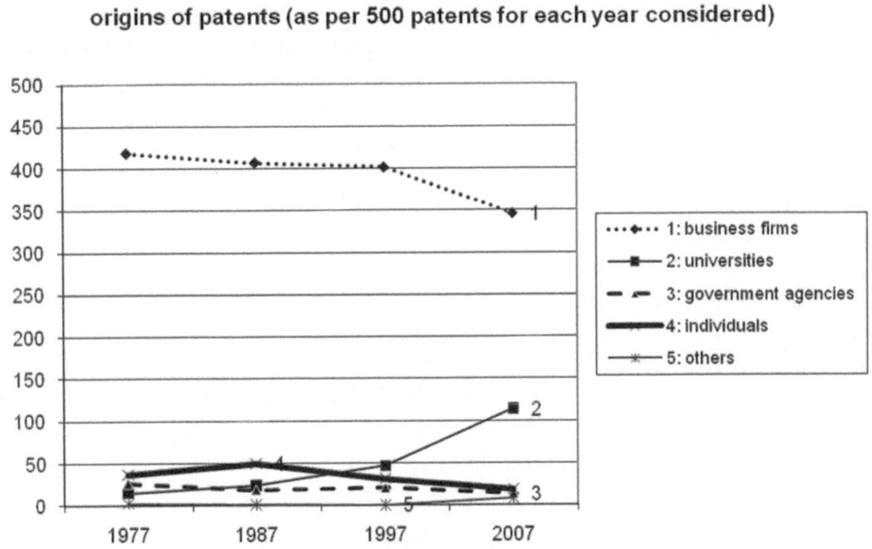

Figure 6.3 Origins of patents, as per 500 patents for each year considered.
Source: Delphion (USPTO, EPO, WIPO); author's calculation.

or less stable, at about 4 percent of the total. These appear in such headings as 'The Secretary of State for Defense in Her Britannic Majesty's Government of the United Kingdom and Northern Ireland' or 'The United States of America as represented by the Department of Health, Education and Welfare.' Some law analysts have raised concerns about how potential liabilities of these patents would be treated, questioning who would be responsible in case of litigations. But the most significant trend is the steady decline of patents granted to business firms, mitigated by the steady increase of patents granted to universities and other research organizations. The increase of the share of patenting granted to universities reflects the fact that nano patenting is more and more the result of group work. Individuals are not in a position to compete with whole teams. Also, universities have gradually moved toward a research policy whereby their researchers surrender to them their patenting rights. The gradual decline of the number of business firms as assignees of nanopatents has another explanation. Firms increasingly consider universities as subcontractors for R&D. They prefer to fund university research departments than to carry in-house research. Again, the resources needed in order to obtain results may explain why it is more efficient for them to do so. Another reason for doing so is that they disconnect their activities from R&D. If new products or processes prove harmful, the universities developing them will be associated with them—not the firm.

Finally, I looked at the evolution of the areas that were granted the most patents over the last thirty years or so in order to have an overview of the most likely fields of entrepreneurial activity. To define these areas, I used IPC codes. The International Patent Classification (IPC) is currently used by over seventy patent authorities to classify and index the subject matter of published patent specifications. The IPC is maintained and administered by the WIPO and was first published in 1968. The most recent edition came into force in January 2006. The IPC divides patentable technology into eight key areas, which are then subdivided into child classes. Some areas overlap, and some patents may fall under several headings. The classification contains one section devoted to nanotechnology. It is section B82, NANOTECHNOLOGY. However, this section has the same characteristic as USPTO Class 977: it contains few patents, and most of the nanopatents that have been granted do not fall under its heading.

Based on calculations made using the Delphion relevancy score, Figure 6.4 illustrates the evolution of patenting areas. The relevancy score is based on algorithms built into Delphion's search engine that factor in the density

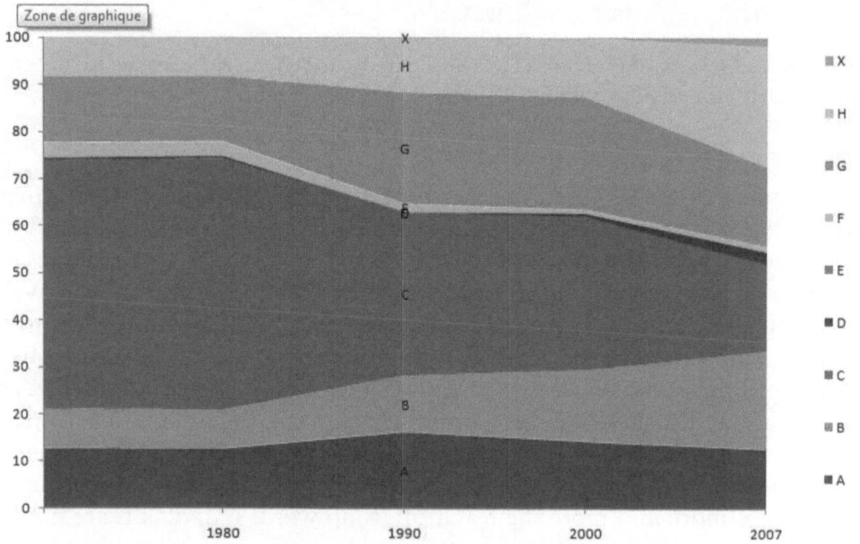

Figure 6.4 Evolution of patenting activity by IPC area.
Source: Author's calculation.
Key:
A: Human Necessities (subdivided into sixteen child classes)
B: Performing Operations, Transporting (subdivided into thirty-seven child classes)
C: Chemistry, Metallurgy (subdivided into twenty-one child classes)
D: Textiles, Paper (subdivided into nine child classes)
E: Fixed Constructions (subdivided into eight child classes)
F: Mechanical Engineering, Lighting, Heating, Weapons (subdivided into eighteen child classes)
G: Physics (subdivided into fourteen child classes)
H: Electricity (subdivided into six child classes)

of the search terms found in the document. Density is calculated in proportion to overall document text. When proximity operators are used, the nearness of terms within the document is also a factor. The relevancy score allows users to work on samples of the five hundred most relevant results.

Figure 6.4 shows that nanopatenting overwhelmingly began in the area of chemistry and metallurgy (classification C), with nearly 53 percent of all nanorelated patents. Over the years, patenting spread out increasingly evenly among five categories: A—Human Necessities (12.6 percent); B—Performing Operations, Transporting (20.8 percent); C—Chemistry, Metallurgy (18.4 percent); G—Physics (16.8 percent); and H—Electricity (25.8 percent). The great 'winners' were classifications B (Performing Operations, Transporting) and H (Electricity), which had moved, respectively, from 8.3 to 20.8 percent, and from 8.3 to 25.8 percent. An interesting phenomenon was the appearance of what I classified as X—that is, patents without any classification code. These did not exist in earlier times, but by 2008 they amounted to 1.6 percent of all the nanopatents, which is not really negligible. One section remained remarkably unaffected, classification E (Fixed Constructions). This may reflect the reality that nanotechnologies are still a long way out from the application stage in this domain. Applications seem more readily accessible in textiles, as illustrated by the fact that classification D started taking off around 2000.

7 Nanoproducts Reaching the Market

Much speculation is going on surrounding predictions of what nanotechnology entrepreneurship will look like in the near future. Several conclusions may be drawn from the analysis of R & D funding and of patent trends as seen in chapter 6. One is that nanotechnology R&D is a global activity that involves three teams of players: the Americas, Asia, and Europe. Apart from them, the other regions of the world are for the moment only outsiders. In the group of three, globalization is felt not only in the exchanges of researchers coming from various parts of the world, but perhaps even more so in the fact that each region monitors carefully what the others are doing and responds accordingly in order to enhance its position. So although there are still important structural differences between approaches, there is an evolution toward a global integration of priorities and, to a certain extent, of road maps. This points to a globalization of the major areas of development for nanotechnology entrepreneurship.

Second, in spite of all the research that is going on to assess nanotechnology R&D, it is very difficult to obtain an accurate picture of the situation. As different regions or organizations use different methodologies and classifications, comparisons are not easy. There is also a growing tendency to include in classifications categories such as 'NEC' (not elsewhere classified), or 'unknown,' in which up to 10 percent of the data is stored, which leads to further difficulties in interpretation, especially when the differences are small. Bearing this in mind, what do the recent studies of nanoproducts reaching the market tell us?

In its nanotech report, Lux Research Inc. (2007) tried to forecast the commercial evolution of the technology. The researchers interviewed more than one hundred executives, researchers, and academics working to commercialize nanotechnology and integrated macroeconomic data from the World Bank and the U.S. Bureau of Economic Analysis, among others, in order to build models for forty-two product segments connecting with nanotechnologies. They found that sales of products incorporating nanotechnologies would rise from nearly nothing in 2004 to 15 percent of the total global manufacturing output in 2014. This value approached the size of the combined information technology and telecom industries, and was ten times larger than the revenues from biotechnologies. The report showed that the economic impact of nanotechnologies would not come from the sale of staple products such as carbon nanotubes, but rather from very diverse products and goods from different industrial sectors that would integrate the technology at some point. Their projection of revenues from products incorporating nanotechnologies was that in 2014, 4 percent of general manufactured goods, 50 percent of electronics and IT products, and 16 percent of goods in health care and life sciences would incorporate nanotechnologies. Lux Research Inc. modeled nanotechnology growth as a three-phase process: after an initial phase in which nanotechnology was incorporated selectively into a few high-end products, the second stage would see an explosion in the sector of electronics and IT applications when microprocessors and memory chips using new nanoscale processes reached the market. This analysis was in line with my own analysis of the evolution of patenting areas (Figure 6.4). Then, toward 2010, according to the same report, there would be a third stage with the generalization of manufactured goods including nanotechnologies and the beginning of nanoenabled pharmaceuticals and medical devices, as trial times would be over for health care and life sciences applications (see Figure 7.1).

Another study of the types of nanoproducts reaching the market came from Credit Suisse, one of the world leaders in financial services. The study was intended to offer guidance to investors, and its conclusions were presented at the Nanotech Northern Europe 2008 conference that took place

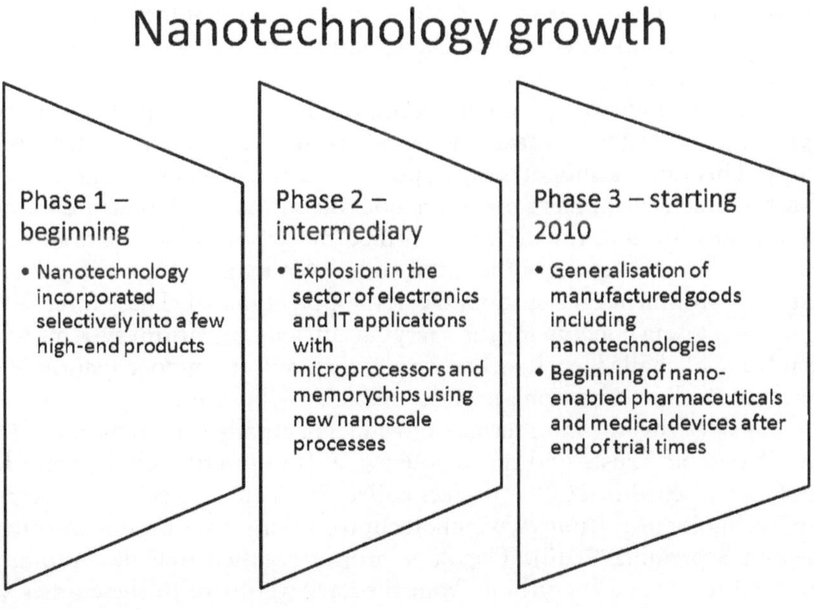

Figure 7.1 Nanotechnology growth.
Source: Lux Research Inc.; adapted.

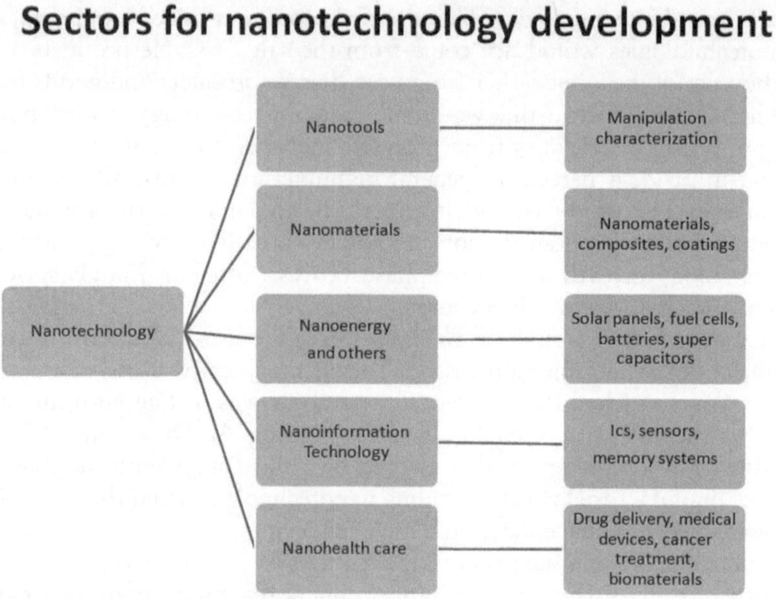

Figure 7.2 Sectors for nanotechnology development.
Source: Credit Suisse.

in Denmark in September 2008 (http://www.nanotech.net). Credit Suisse identified five major areas of nanotechnology development: nanotools, nanomaterials, nanoenergy and others, nano information technology, and nano health care (Figure 7.2).

Credit Suisse identified "population growth, better life quality, and rising energy and water and security needs" as the main drivers of nanotechnology. This finding supports my hypothesis that awareness of the concept of sustainable development plays a major role in the development of nanotechnologies. Indeed, the drivers identified by Credit Suisse relate directly to a pattern of resource use that aims to meet human needs while preserving the environment on a sustainable development model. These forces also tie in a prevailing concern for the carrying capacity of natural systems with today's social challenges. In other words, they reflect a preoccupation with environmental preservation and sustainable development.

What does this mean in terms of products actually on the market? The Pew Charitable Trusts and the Woodrow Wilson International Center for Scholars launched in 2005 a project called PEN, the Project on Emerging Nanotechnologies (http://www.nanotechproject.org/inventories/consumer/, accessed September 2010). The PEN project recalled that nanoproducts promised to change everything from the cars we drove to the clothes we wore, from the medical treatments our doctors could offer to our energy sources and workplaces. The list of goods and services to come—from

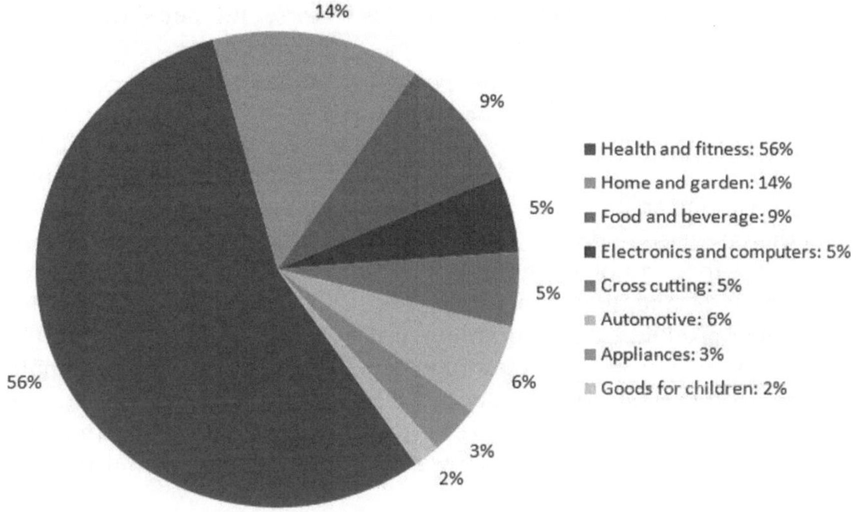

Figure 7.3 Nanoproducts on the market by category in August 2010.
Source: PEN. Author's chart.

new cancer therapies to pollution-eating compounds, from more durable consumer products to detectors of such biohazards as anthrax, from novel foods to more efficient solar cells—illustrated the breadth of potential business opportunities. Part of the project's goal was to provide an online inventory of all nanotechnology-based consumer products currently on the market. The inventory dated August 9, 2010, listed 1,090 different products produced by 485 companies located in 24 countries worldwide. The inventory split along eight categories: health and fitness, home and garden, electronics and computers, food and beverage, cross cutting, automotive, appliances, and goods for children.

Figure 7.3 shows how significant the category health and fitness was in relation to the others. This category contains several subcategories, in which cosmetics, clothing, and personal care dominate.

Finally, more recent studies, including Zhu (2008), identified construction as an extremely promising sector for nanotechnology applications. Zhu explained that the construction industry was among the earliest—in the 1990s—to categorize nanotechnology as a promising emerging technology. Yet, because of a tradition of low R&D investment in construction and of an absence of communication between scientific circles and the construction industry, the rate of application stayed very low until 2002–3. Since then, research on cement and concrete has received a great deal of attention, with environmental and sustainability concerns identified as major drivers of progress. Although nanotechnology applications in construction still lag behind other industries, commercial activities have begun to emerge in the high-performance and functional materials sectors, and they show an enormous potential for development.

As to the raw materials themselves, data from the French Agency for Environmental and Occupational Health Safety (AFSSET 2010 [renamed ANSES in 2010]) and from the PEN inventory of consumer nanoproducts indicate that the major materials in terms of production are silver, followed by carbon and titanium. At the nano size, silver presents antibacterial properties that are used in deodorants as well as in functionalized fabrics (anti-odor socks, for example). Nanocarbon is popular with tire makers, tennis rackets makers, and many other industries seeking extra strength and duration in their products. As for titanium, it is widely used in the form of nano titanium dioxide in the cosmetics industry. These are only a few examples. Because of the versatility introduced by the simple change of scale, either from bulk size to nanometric size or even between different nanometric sizes, each of the materials can have many other properties and therefore find many other applications.

8 Summary

The purpose of part two was to provide a broad outline of nanotechnology entrepreneurship by focusing on a few representative indicators, which would define the context from which to draw observations concerning the characteristics of institutional emergence and change accompanying it. At the time of this writing, nanotechnology research mobilized only a *very* small fraction of the world's GDP. This means that whatever findings the study of nanotechnolgy entrepreneurship indicators could produce could only be inductive in essence, vindicating the methodological approach of this book, which is to build theory from the observation of a few illustrative cases.

The fact that the early development of nanotechnology corresponded to the creation of ad hoc coordinating structures such as the American NNI or the European Nanoforum shows the importance of policy making in both the institutional and the future industrial emergence of the field. The importance of policy making is also illustrated by the fact that European Framework Programmes FP6 and FP7—when European nanotechnology R&D funding really took off—were the largest ever funding programs for research and technological development (RTD). The shift in the patenting model toward the domination of university patenting for nanotechnology discoveries is a further indication of the importance of policy making, as university research tends to correlate to institutional funding by research councils.

The analysis of the categories of nanoproducts reaching the market as well as of the areas of growth for the technology illustrates the growing importance of sustainable development issues in the development of nanotechnologies.

Part III

Nanodevelopment and Regulatory Practices

9 Areas of Development

In the previous part, I began my exploration of the emergence of the field of nanotechnologies with an outline of nanotechnology entrepreneurship. It was based on three indicators: nanotechnology R&D, nanotechnology research centers and infrastructures and, finally, nanorelated patent activity. Among other findings, I revealed that institutional funding and R&D orientations play an important role in nanotechnology entrepreneurship, and that sustainable development was one of the drivers of nanotechnology entrepreneurship. In this section, I focus on regulatory practices. My objective in the next pages is to demonstrate that regulatory practices play a vital role in the development of nanobased activities. My longer-term goal is to show that change in regulatory practices signals change in institutional emergence: on the one hand, nanotechnologies are creating a context that is conducive to regulatory change, and at the same time, change in regulatory practices influences the shaping of the field through its impact on nanotechnology entrepreneurship.

With this double objective in mind, I analyze how the regulatory contexts of the nanotech sector in the U.S. and in Europe initially influenced the development of nanobased activities and entrepreneurship differently. The study is based on an interpretative comparative-content analysis of the emergence of the biotechnology and nanotechnology sectors in the U.S. and in Europe. In the U.S., the data relate essentially to federal regulation, and in Europe, to community regulation. I view regulation as the ensemble of official texts and decisions produced by governing structures to set conditions on the development of biotechnologies and nanotechnologies. These data are complemented by focused interviews with officials working in regulatory agencies, and by business reports concerning the evolution of the biotechnology and nanotechnology sectors. The analysis is structured as follows: I first recall the major areas of development of nanotechnologies. I then describe the specifics of the evolution of regulations designed to accompany the development of technological discoveries and their impact on society. Next I discuss the robustness of the different regulatory frameworks to exogenous factors. Drawing on a comparison with a longitudinal case study of biotechnology, I discuss the influence of regulatory contexts on entrepreneurship. I conclude that over the period of the early emergence of the nanotechnology field (2001–6), in spite of a perceivable trend toward globalization and harmonization of regulatory frameworks in general, Europe and the U.S. were not aligned when it came to nanotechnology regulation, because each regulatory framework created its own competitive advantage for entrepreneurship.

As has already been discussed, *nanotechnology* is a generic word that refers to products and technologies involving the manipulation of matter at the level of atoms on the scale of the nanometer (nm), i.e., a millionth of a millimeter. Over the last ten years, there have been countless arguments over the exact definition of nanotechnology, especially as particles of the nanometer size have been around as long as the world has been in existence. For example, there have been carbon nanoparticles in soot and volcano exhausts since the beginning of time. It is now agreed that the field of nanotechnology concerns man-made particles of a nano size that present properties that were not previously present in a natural state. Again, to be counted as a particle of nano size, it is generally agreed that at least one of a particle's dimensions must be inferior or equal to 100 nanometers (nm). Manipulating matter at such a scale, by acting directly on the atomic structure of particles, seems to open up endless possibilities. Today, nanoproducts that are on the market—i.e., products that contain specific man-made nanoparticles—include self-cleaning glass or textiles, reinforced tires or bumpers, plasters or refrigerators containing nanosilver particles to kill bacteria, cosmetics, toothpaste, cements or paints with specific properties, food, beverages, wrappings, fuel additives, to name just a few. In terms of the products of research, nanotechnology is at least as full of promise for entrepreneurs as biotechnology was in the 1970s. There are currently three major identified fields of development for all the products and techniques obtained by working on nanoparticles: the health sector, military applications, and sustainable development through the development of new or better products in electronics, fuel additives, or industrial materials.

NANOTECHNOLOGY AND HUMAN-HEALTH APPLICATIONS

The health sector is the best-known sector of development and entrepreneurship involving nanotechnology applications. It is amply documented, particularly as most developments of nanotechnology research in the health sector intersect with biotechnology research. (See, for example, Ernst & Young 2003; U.S. Department of Commerce 2003; Vo-Dinh 2007).

The following is a brief summary of the most promising applications of the use of nanotechnologies for human health:

DNA-*based*

Gene probes, DNA markers, bioinformatics, genomics, pharmacogenetics, DNA sequencing/synthesis/amplification, genetic engineering.

Biochemistry/Immunology

Vaccines/immune stimulants, drug design & delivery, diagnostic tests, antibiotics, synthesis/sequencing of proteins and peptides, cell receptors/signaling, structural biology, combinatorial chemistry, 3-D molecular modeling, biomaterials, microbiology, virology, microbial ecology.

Bioprocessing-based

Culturing/manipulation of cells, tissues, embryos, extractions, purifications, separations, fermentation, bioprocessing, biotransformation.

Environmental

Bioleaching, biopulping, biobleaching, biodesulfurization, bioremediation, biofiltration.
Source: U.S. Department of Commerce Technology Administration, Bureau of Industry and Security.

NANOTECHNOLOGY AND THE MILITARY SECTOR

It is more difficult to find comprehensive statistical data concerning the development of nanotechnologies in the military sector. The Department of Defense (DOD) accounted for about a third of the yearly budget that the U.S. Congress had appropriated for FY 2003–5 for nanotechnology R&D across all the agencies participating in the NNI (National Nanotechnology Initiative). For the period 2004–6, the NNI changed its budget presentation around the notion of Program Component Areas (PCAs). This made it harder to single out how much of the general budget the U.S. Congress appropriated to Defense for nanotechnology R&D, but the DOD appeared to count for about a quarter of the investment in nanotechnology research in the NNI in 2006 (U.S. Department of Defense 2005; NNI 2007). Several new institutes focused on military applications, including the Institute for Soldiering Nanotechnology at the Massachusetts Institute of Technology; the Center for Nanoscience Innovation for Defense at the University of California, Santa Barbara; and the Nanoscience Research Laboratory of the Naval Research Laboratory, which is funded with Military Construction (MILCON) funds.

In Europe, MINATEC was in 2008 the leading European nanotechnology research center. It was based in Grenoble, France, but was born from the association of several partners. Among the major partners were the French DGA (Délégation Générale à l'Armement) and CEA (Commissariat à l'Energie Atomique), and according to a report published by CROCIS (Centre Régional d'Observation du Commerce, de l'Industrie et des Services), nanotechnology public research investments in France focused more on the military and energy sectors than on the health sector (Rigaud 2003).

The first plant dedicated to the production of nanoparticles in the UK was inaugurated in 2003 in Farnborough in Hampshire, and belonged to the Qineti Group. QinetiQ was created in 2001 as a public-private partnership (PPP) takeover of part of the UK Ministry of Defence (the Defence Evaluation and Research Agency—DERA), and the Ministry of Defense was its largest customer. It became one of the world's leading defense technology and security companies. QinetiQ Nanomaterials Ltd, one of its spinout companies, became the largest nanometric materials and nanotechnology group in Europe (Source: www.qinetiq.com, accessed August 10, 2008). The proceedings from a workshop on nanoscience for the soldier sponsored by the U.S. Army Research Office summarize the dimensions explored for developing new products and services in this area (North Carolina Biotechnology Center 2001).

Materials and Fabric

High-strength, ultralightweight material, adaptive multifunctional materials

Power, Energy Distribution, and Cooling

Revolutionary power sources to enable soldiers to operate autonomously for up to 144 hours (i.e., six days); warrior body-cooling devices to enable sustained performance under incapacitating environmental conditions (extreme heat, extreme cold, polluted or poisoned atmosphere).

Soldier Status Monitoring and Modeling

Soldier status monitoring (operational capability, NBC—nuclear, biological, and chemical warfare—detection, casualty monitoring); soldier performance prediction and enhancement; soldier casualty minimization.

Displays, Detectors and Antennas

Communications between "suit" and helmet/soldier; information management; security of soldier system.
Source: U.S. Army Research Office.

NANOTECHNOLOGY AND SUSTAINABLE DEVELOPMENT

As for the field of sustainable development, it may end up becoming the leading entrepreneurial sector, as the necessity to find new energy sources and to preserve the environment becomes more and more a priority. The United Nations Environmental Program (UNEP) published in 2007 a special section devoted to nanotechnologies in its yearly Global Environmental Outlook—GEO (UNEP 2007). A Canadian study released in April 2005 (Salamanca-Buentello et al. 2005) focused on the potential use of nanotechnology to address some of the world's most critical development problems, and identified the ten applications of nanotechnology most likely to benefit developing countries, while demonstrating that these applications could contribute to the attainment of the United Nations Millennium Development Goals (MDGs). The study also revealed that developing countries were already harnessing nanotechnology to address some of their most pressing needs. Although the study was done to account for the needs and development potential of developing countries, it could also be applied to developed countries faced with sustainability issues. The ten applications of nanotechnology most likely to benefit developing countries were as follows:

1. Energy storage, production, and conversion
2. Agricultural productivity enhancement
3. Water treatment and remediation
4. Disease diagnosis and screening
5. Drug delivery systems
6. Food processing and storage
7. Air pollution and remediation
8. Construction
9. Health monitoring
10. Vector and pest detection and control

The study included references to the health sector, which shows how closely interconnected the potential developments of nanotechnology applications are. Concrete actions are following. For example, Argentina's National Agency for the Promotion of Science and Technology (ANPCYT) earmarked 15 percent of its 2008 budget for scientific and technological research to nanomaterials and nanodevices for the promotion of agriculture and food production (UNESCO 2010). The 2010 UNESCO science report noted that under the leadership of Brazil, Argentina, and Mexico, Latin America was becoming a world specialist in the field of nanomedicine and nanobiology, according to records of patent development and ownership.

10 Initial Positive Regulation

To discuss the influence of regulatory contexts on the development of the nanotech sector, I draw on a comparative analysis of the development of biotechnology. I take as the origin of my study (like the origin of a graph) the discovery of the three-dimensional double-helix structure of DNA by James Watson and Francis Crick in 1953. Although research is a continuous process, and this discovery could not have happened without the previous research that led to it, it is generally considered to be the starting point of the emerging field of biotechnology. At this point of origin, the stakeholders were the scientists involved in the discovery and their supporting institutions. Watson was American and Crick was British, but the discovery was made in England, at Cambridge University. When Watson and Crick discovered the double-helix structure of DNA, they were not entrepreneurs; they were scientists. However, they quickly became entrepreneurs—as opposed to other scientists working in labs who did not take the same course of action, they moved on to new structures that would connect them to the market. As it happened, they both found these structures in the U.S.: Crick did brain research at the Salk Institute for Biological Studies in La Jolla, California, and Watson became director of the molecular-biology lab at Cold Spring Harbor, New York, and served as the head of the Human Genome Project.

The attributes that most attracted them to their new institutions were regulation and funding, with the two being very closely related. Firstly, American institutions clearly favored an absence of restrictive regulation. In 1977, although sixteen bills had been introduced in the American Congress to regulate recombinant DNA research—particularly to prevent escapes of the developed products from the laboratory to the outside environment—not one was passed. At the same time, the UK was one of the first countries to introduce regulation concerning research on genetic modifications: in 1976, the UK Medical Research Council set up the Genetic Manipulation Advisory Group (GMAG), and the 1978 Health and Safety Regulations requested notification of any activity involving genetic manipulation.

In addition to an absence of restrictive regulation, the U.S. also developed 'positive regulation.' Positive regulation is to be understood as regulation that encourages the development of an emerging field, as opposed to "restrictive regulation," which is often seen as putting restraints on the development of technological innovation. Positive regulation took several forms in the U.S. For example, in 1980 the U.S. Supreme Court ruled that genetically altered life forms can be patented and granted scientist Ananda Chakrabarty a patent for oil-eating bacteria. This ruling opened up enormous possibilities for the commercial exploitation of genetic engineering, as patents enabled companies to begin cashing in on laboratory discoveries. This was positive regulation. Another piece of positive regulation came with the Bayh-Dole Act of 1980, which gave universities and nonprofit institutions control of the inventions that arose from research supported by federal funding.

At that stage, regulation and funding were closely interrelated: more than 60 percent of the Salk Institute for Biological Studies' research was funded by U.S. government grants. Thanks to the Bayh-Dole Act, and with the help of the Salk's Office of Technology Management (OTM)—which was created in the late 1960s to maximize patent protection and to facilitate the transfer of the Institute's technology to the market—technology management at the Salk Institute resulted in over 350 U.S. patents and more than 250 license agreements with pharmaceutical, biotechnology, and reagent supply companies. In addition, seventeen start-up companies were founded to develop technology for the institute. With such an enabling environment, the Salk Institute was an ideal place to foster academic technology transfer, and this was one example among many where research done at academic institutes led to new business ventures. Hundreds of companies and tens of thousands of jobs were thus created in the San Diego region alone. Finally, regulatory support was boosted by political publicity: in 1981, Congressman Al Gore held a series of hearings on the relationship between academia and commercialization in the arena of biomedical research, focusing on the effect that the potential for huge profits from intellectual property and patent rights could have on the research environment at universities.

Although Jonathan King, a professor at MIT speaking at the Gore hearings, reminded the biotech industry that the most important long-term goal of biomedical research was to discover the causes of disease in order to prevent disease, (Ruscio 1994) from the early start the prospect of huge economic returns was a key motivation for entrepreneurship. For example, that same year, Hoechst AG, the West German chemical company, gave Massachusetts General Hospital, a teaching facility of Harvard Medical School, $70 million to build a new Department of Molecular Biology in return for exclusive rights to any patent licenses that might emerge from the facility. The Human Genome Project is another example of how the combination of regulation and funding led to entrepreneurship. It began formally in 1990 as an effort coordinated by the U.S. DOE and the NIH to map the position of every human gene and to read and decipher every message encoded in the twisted double helix of human DNA. It was originally planned to last fifteen years, but rapid technological advances accelerated the completion date to 2003. The project later became truly international, but it nevertheless benefited from several hundred million dollars coming from the DOE and NIH.[1] This huge investment of public money went hand in hand with an enormous amount of knowledge transfer and business creation. In the genomics subfield alone, dozens of small companies sprung up to sell information, technologies, and services to facilitate basic research into genes and their functions.

To encourage the early development of nanotechnologies, the U.S. followed the same model mixing funding and positive regulation, but added a new step to this strategy: integration. In 2001, the government created a single program centralizing the major research efforts relating to nanotechnology: the NNI (National Nanotechnology Initiative). The NNI's website stated that the initiative was a federal R&D program established to coordinate the multiagency efforts in nanoscale science, engineering, and technology in order to maintain a world-class research and development program, to facilitate technological transfer, to develop educational resources and to support responsible development of nanotechnology.

As already discussed in previous chapters, the NNI gathered at its birth in 2001 six federal agencies. In six years, this number rose to twenty-five federal agencies, among which thirteen had a special nanotechnology R&D budget. During the same period, spending nearly tripled, from an estimated $464 million in 2001 to $1.4 billion in 2007. The U.S. centralization promoted by the federal government was instrumental in making the whole sector more efficient and in promoting entrepreneurship. The report published in May 2005 by the U.S. President's Council of Advisors on Science and Technology (PCAST 2005) found that the U.S. was the world leader in nanotechnology as well as the acknowledged leader in nanotechnology research and development. The approximately $1 billion the federal government would spend on nanotechnology R&D in 2005 was roughly one-quarter of the current global investment by all nations. Total annual U.S. R&D spending (federal, state, and private) stood at approximately $3 billion, or one-third of the estimated $9 billion in total worldwide spending by the public and private sectors combined. In addition, the United States led in the number of start-up companies based on nanotechnology and in research output as measured by patents and publications. The report concluded that the NNI appeared well positioned to maintain U.S. leadership going forward, thanks to its coordinated interagency approach to planning and implementing a federal R&D program. As a result, PCAST recommended continued robust funding for the NNI.

In Europe, nanotechnology only took off with programs covering all European research in 2004, with Framework Programme 6 (FP6). Before that, the European Commission had launched in 2000 the European nanotechnology network Phantoms. It was a scheme intended to promote European science and research through a plurinational networking action putting together research capacities present in various European regions and stimulating commercial nanoelectronic applications. The Phantoms network was of a truly interdisciplinary character and included members coming from government, universities, and industry. In 2004, the network included 237 partners from 32 countries and gathered approximately 2,100 researchers. However the partners did not all come from Europe. Eight partners were from the U.S., two from Canada, and one each from Argentina, Australia, Cuba, India, Israel, Japan, and Venezuela. The network was essentially a virtual structure of coordination and exchange of information.

In February 2004, the European Commission founded a new structure called Nano2Life, based in Grenoble, France (see http://www.nano2life.org). Its objective was to merge existing European expertise and knowledge in the field of nanobiotechnology. Nano2Life was tackling the fragmentation of European nanobiotechnology by joining twenty-three previously unconnected, highly specialized regions and centers with experience in initiating and running nanobiotech programs. The network included a pool of forty-one high-tech companies. Nano2Life aimed at creating the foundation for a virtual European nanobiotech institute focused on the understanding of the nanoscale interface between biological and nonbiological entities and its possible application in the area of complex and integrated novel sensor technologies for health care, pharmaceuticals, environment, security, food safety, etc. However, there was still in 2008 no strict equivalent to the U.S. NNI, even if coordination was well under way.

The overall structure coordinating European nanotechnology was Nanoforum, the European Nanotechnology Gateway (see http://www.nanoforum.org). Nanoforum is a virtual network that provides a comprehensive source of information in all areas of nanotechnology. It was originally funded by the European Commission through FP5 in 2003. Since July 2007, it has been operating as a European Economic Interest Grouping (EEIG), giving it legal status and the ability to enter into contracts with other parties, including tenders and new EU-funded projects.

In the early development of the nanotechnology sector, it was clear that the European lack of coordination was detrimental to entrepreneurship. According to a survey carried out by Nanoforum (Nanoforum, Malsch, and Oud 2004), North America was perceived to be the world leader both in nanosciences (76 percent of the respondents) and in the transfer of nanotechnology to industry (77 percent of the respondents), with Europe and Asia falling far behind. This shortfall was mainly attributed to a lack of a coherent system of infrastructure, with the majority of respondents (64 percent) highlighting the need for a new large infrastructure at the European level.

One may summarize the above by saying that just as it had played a key role in favoring entrepreneurship in the biotech sector, positive regulation was instrumental in giving the U.S. a head start in the early emergence of the nanotechnology sector. Positive regulation allocated money—vast sums of it—to an emerging sector, and it immediately aimed for the integration of the sector with the creation of a dedicated program (the NNI), which were two of the most important aspects of positive regulation. The creation of this centralized institution preceded the writing of an act defining what was going to be done. This was done later thanks to two texts. The first one, *National Nanotechnology Initiative: Leading to the Next Industrial Revolution,* emanated from the U.S. White House on January 21, 2000, and was a four-page press release with the following introduction:

> President Clinton's FY 2001 budget request includes a $227 million (84 percent) increase in the government's investment in nanotechnology research and development. The Administration is making this major new initiative, called the National Nanotechnology Initiative (NNI), a top priority. (White House 2000)

Then followed a brief summary of the potential breakthroughs nanotechnology might lead to, along with an outline of the funding strategy. The second text, Act S 189, also called *21st Century Nanotechnology Research and Development Act*, was a congressional act of ten pages explaining the functioning of a national nanotechnology program, which in other words was the functioning of the NNI, with a clearly dated and budgeted road map (U.S. Congress 2003).

In Europe, the process was in a sense reversed. In 2005, two major defining texts had been written: *Towards a European Strategy for Nanotechnology* and *Nanosciences and Nanotechnologies: An Action Plan for Europe 2005–2009* (European Commission 2004, 2005). Yet Nano2Life (set up in February 2004) was still only the preparatory structure of a future virtual European nanobiotech institute, and there was no single integrated European nanotechnology agency. The tables of contents of the two European reports summarize the differences from the U.S. approach. First, the EU texts were much longer than the U.S. ones. Then, unlike the U.S. texts, the European ones had no performative value in the sense that they did not *create* institutions. They merely described and defined what was to become the field of nanotechnologies as conceived by the EU, with specific parts on safety, the third world, and the necessity to integrate energies (Figures 10.1 and 10.2). Figure 10.3 summarizes what differentiates the two approaches.

Towards a European Strategy for nanotechnology (28 pages)

Figure 10.1 Table of contents of *Towards a European Strategy for Nanotechnology. Source: European Commission.*

Nanosciences and Nanotechnologies:
An action plan for Europe 2005-2009 (12 pages)

Figure 10.2 Table of contents of *Nanosciences and Nanotechnologies: An Action Plan for Europe 2005–2009.*
Source: European Commission.

Two different visions of nanotechnology positive regulation

US: an act setting up an integrated structure

- *National Nanotechnology Initiative: Leading to the next industrial revolution.* US White House, 21 Jan 2000
- *21st Century Nanotechnology Research and Development Act.* S. 189. 108th Congress, 1st Session, 18 Nov 2003

EU: a strategic plan framing a field

- *Towards a European Strategy for Nanotechnology.* COM(2004) 338, European Commission, 12th May 2004
- *Nanosciences and nanotechnologies: An action plan for Europe 2005-2009.* COM (2005) 243, European Commission, 7th June 2005

Figure 10.3 Two different visions of nanotechnology positive regulation.
Source: author.

11 Initial Restrictive Regulation

At the end of the first decade of the twenty-first century, U.S. federal regulation was gathered in the *United States Code*, the register of all U.S. federal regulation. In Europe, community regulation was gathered in the *Directory of Community Legislation* (Government Printing Office 2009; European Communities 2010). In the U.S. as in Europe, this regulation constituted the minimum requirement that must be respected by all the parties (the states in the U.S., and the various members of the EU). This meant that individual states in the U.S. and individual EU members in the EU *might* have additional regulations if they wished, but *must* comply with the existing regulation. In the EU, the acceding member states (in 2011, the EU consisted of twenty-seven member states) had the obligation to fully and correctly transpose and implement the 'acquis communautaire' (the EU *acquis*) by the date of accession so as to guarantee equivalent rights and obligations to all citizens of the enlarged European Union. The 'acquis communautaire' amounted to some 1,600 directives. Beginning from the date of accession, the acceding countries were monitored in the same way as the previous member states, and monitoring and compliance updates were published at regular intervals (European Commission 2004). Eur-Lex, the portal to European Union law (http://europa.eu.int/eur-lex/en/index.html) also provided studies and statistics concerning the application of community law in member states. In the U.S., individual states must follow federal regulation more or less in the same manner and, on the whole, both systems worked in a fairly similar fashion.

However, the contents of regulation were quite dissimilar. One element of disparity was the concept of 'precautionary principle.' The first EU mention of the 'precautionary principle' was in the environment title of the Treaty of Maastricht establishing the European Community in 1992 (European Community 1992). At that stage, the notion was mentioned in the context of marine pollution and only referred to, not defined:

> 2. Community policy on the environment shall aim at a high level of protection taking into account the diversity of situations in the various regions of the Community. It shall be based on the precautionary

> principle and on the principles that preventive action should be taken, that environmental damage should as a priority be rectified at source and that the polluter should pay . . .
> *Source:* The Maastricht Treaty, Title XVI, Environment, art. 130r, 2, p. 37.

In 2000, the EU wrote a complete text of reference concerning the precautionary principle called the *Communication from the Commission on the Precautionary Principle* (Commission of the European Communities 2000). This then served as a basis for all EU regulation; there was no counterpart in U.S. legislation. The most important area of variance—and disagreement—with U.S. law circled around the following statement:

> The absence of scientific proof of the existence of a cause-effect relationship, a quantifiable dose/response relationship or a quantitative evaluation of the probability of the emergence of adverse effects following exposure should not be used to justify inaction.
> *Source:* Commission of the European Communities, *Communication from the Commission on the Precautionary Principle* (2000), p. 17.

In essence, what this rather confusing statement meant was that action could be taken against some specific products, even in the absence of *proof* that they were dangerous. The U.S. rejected the precautionary principle, considering that until something was proven to be harmful, it was not to be considered so. The U.S. Chamber of Commerce specifically stated that it opposed the adoption of the precautionary principle as the basis for regulation.

U.S. Chamber of Commerce Strategy:

- Support a science-based approach to risk management, whereby risk is assessed based on scientifically sound and technically rigorous standards.
- Oppose the domestic and international adoption of the precautionary principle as a basis for regulatory decision making.
- Educate consumers, businesses, and federal policy makers about the implications of the precautionary principle.

Source: U.S. Chamber of Commerce, *Regulatory Reform Issues, Precautionary Principle* (updated September 5, 2005). See also U.S. submission to the Codex Alimentarius meeting in Paris April 10–14, 2000, addressing U.S. concerns over EU communication to the Codex on 'precautionary principle.'

12 Regulatory Frameworks and Exogenous Factors

Attempting to explain why there are different economic or political patterns in different locations has been at the heart of much scholarly research at least since the nineteenth century and Tocqueville's *De la Démocratie en Amérique* (Kagan 2004; Tocqueville [1835] 1999; Kotkin 1994). Most comparative studies have drawn from cultural factors to explain differences observed in various economic systems or, more recently, to account for dissimilarities between national attitudes toward economic power, freedom of trade, or regulatory matters (Kohler 2008; Pape 1999).

According to Pape (1999), in spite of differences between various countries, Europe has on the whole traditionally supplemented market economy with a social component, whereas the American form of Anglo-Saxon capitalism is marked by a strong Darwinian pioneer spirit, as well as by a high level of antagonism between the private and public spheres in general.

As a result, substantial differences exist between U.S. and European laws. These differences become obvious in the way exogenous factors are taken into account in rule making and in the emergence of regulatory frameworks. I will begin by clarifying the concept of exogenous factors. Regulation usually results from the actions of official bodies working according to a democratically accepted procedure. I call 'external stakeholders' individuals or groups of individuals who do not connect to the area of rule making by any lucrative activity. Scientists, policy makers, regulatory bodies are all 'inside stakeholders' because at least part of their revenue comes from their involvement in the area covered by the regulatory framework. The connection between exterior stakeholders and the rule making area is a nonprofit one. These stakeholders have no commercial connection with the area. The forces they represent may take the forms of lobbies, pressure groups, community representatives, and so on. I call their actions 'exogenous factors.'

There is a wide gap between the U.S. and the EU in terms of the robustness of regulation to exogenous factors. Exogenous factors were increasingly part of the lawmaking process in EU regulation, and their 'voice' was heard at an initial stage: when, in 1978, the United Kingdom was the first country in the world to pass legislation controlling genetic modification, the initial thrust had come from scientists themselves (Berg et al. 1975).

These were inside stakeholders. But the restrictions they obtained were quickly eased, and research carried on. However, when in October 1990 all of the EU nations adopted two European Council Directives governing the 'contained use' and 'deliberate release' of genetically modified organisms (GMOs) into the environment,[1] these EU regulations came as answers to growing public concern initiated and relayed by NGOs.

We find the same permeability to exogenous factors in the early talks on the regulation of the nanotechnology sector. In June 2003 in Brussels, a seminar discussing the current state of nanotechnology, examining its risks and promises and proposing steps toward regulating the technology, was jointly sponsored by the Greens/EFA in the European Parliament and by several NGOs (ETC Group, Greenpeace, Dag Hammarskjold Foundation, Genewatch UK, Clean Production Action). The report from the European Commission, *Nanotechnologies: A Preliminary Risk Analysis on the Basis of a Workshop Organized in Brussels on 1–2 March 2004 by the Health and Consumer Protection Directorate General of the European Commission,* was based on a consultation that involved approximately twenty nanotechnology specialists—mostly scientists—but that also included the chief scientist of Greenpeace UK, an environmental group. The report concluded that the behavior of nanoparticles deviated from that of matter at the macroscale, with sometimes diverse effects on environment and health, and that the existing regulations did not distinguish between the differences. It stated that there might be a need for new regulation, and that the precautionary principle suggested avoiding the release of nanoparticles in the environment.

Until the end of the first decade of the twenty-first century, however, the U.S., eschewing exterior stakeholders' views, aimed for *voluntary* nanotechnology regulation (Forrest 1989; Reynolds 2001, 2002; Wardak 2003, 2004). This was confirmed by an August 2005 article from *The Scientist* (Pitman 2005):

> The Environmental Protection Agency (EPA) is spearheading moves to voluntarily regulate the introduction of nanoscale materials in a variety of applications, including the cosmetics and toiletries market.

Essentially, in the areas of the environment and public health, there are four major regulatory agencies in the U.S.: the Consumer Product Safety Commission (CPSC), the Environmental Protection Agency (EPA), the Food and Drug Administration (FDA), and the Occupational Safety and Health Administration (OSHA). There may be other agencies that can impact the environment and public health, such as the Federal Communications Commission and the Department of Agriculture, but they do not have an immediately foreseeable role in nanotechnology. These four agencies have direct regulatory or standards development power. They have the authority to prohibit substances, food, drugs, products, and/or workplace practices. These regulatory agencies (CPSC, EPA, FDA, and OSHA) all provide "notice(s) of proposed rulemaking." The rulemaking is the regulatory action to be undertaken on a product, and it is printed in *The Federal Register.* The notice is an advance warning. In a span of sixty days, typically, any concerned party outside of the regulatory agency can provide input into the decision. If enough public comment is given, a public hearing on the proposed rulemaking is called. Once the sixty-day period has passed, the rulemaking is again published in *The Federal Register,* and the ruling is considered final. The rulings can be found in either *The Federal Register* or the Code of Federal Regulations (CFR).

These regulatory bodies are quite impervious to exterior stakeholders, as is illustrated by the case of titanium dioxide. As sunscreens' active ingredients have evolved, the regulations have changed in kind, and there have been numerous issues with sunscreens for the last twenty-six years. One of the most significant rulings took place at the end of the 1990s. In 1999, the FDA ruled that micronized titanium dioxide was not a new substance relative to its larger particle size counterpart.[2] With this ruling, the FDA denied the existence of nanomaterials as distinct from their macroparticle equivalents. This meant that in spite of the different properties of the molecules at the nano or microscale, the companies that marketed the products using nanoscale titanium dioxide did not need to proceed to any specific toxicity test because the molecules were not considered new chemicals. Although it received public comment in support of considering micronized titanium dioxide a new ingredient, the FDA, "[did] not consider micronized titanium dioxide to be a new ingredient but consider[ed] it a specific grade of the titanium dioxide originally reviewed by the Panel."[3]

Another sign of the U.S. regulatory system's strictness toward exterior stakeholders was a resolution that was introduced in the U.S. Senate in March 2005:

> Expressing the sense of the Senate that judicial determinations regarding the meaning of the Constitution of the United States *should not be based on judgments, laws, or pronouncements of foreign institutions*. . . (109th Cong., 1st Sess., S. Res. 92 (March 20, 2005; italics added).

This statement clearly indicated that the U.S. did not intend to modify its regulatory system to integrate a 'foreign perspective'—essentially another form of exterior stakeholders.

13 Sustainable Development and Nanobased Business Activities

In a paper exploring the variation in rates of entrepreneurship in the United States from 1899 to 1988, Shane (1996) expanded on the work of Schumpeter (1939, 1947) to explain that it is entrepreneurs, rather than managers in established firms, who are recognizing the opportunities for most of the new production functions. His reasoning was that new technological developments often differ so much from the way current technology works that they cannot be incorporated into the technological framework of existing firms. The conflict between the demands of the new technology and the production function of the established firm means that managers in established firms are less likely than entrepreneurs—who are not reliant on the existing production function—to accept the new technology. With new technologies such as nanotechnologies, entrepreneurship tends to happen at the level of institutions or organizations like university spinouts or corporate start-ups because this is where one can find the vital combination of research and initial investments. Yet the same model is followed: entrepreneurs are not reliant on existing production functions. So when biotech or nanotech entrepreneurs want to utilize the opportunities created by scientific inventions to produce new combinations of production factors, they look for the most favorable environment in which to transform these opportunities into profitable business activities.

This explains why the U.S. had a head start in the development of new business activities in the biotech and nanotech sectors. Entrepreneurs go where the terrain is most favorable and, at least initially, U.S. regulation rendered the terrain more favorable. From the outset, first with the biotech sector, and then with the emergence of the nanotech sector, the U.S. provided a mixture of positive regulation and an absence of a regulatory framework, which fostered entrepreneurship in these sectors.

However, entrepreneurship in new technologies is volatile. Entrepreneurs learn very quickly to always try to place themselves in the environment most favorable to generating new institutions creating future goods and services around an emerging technological field. The EC laws concerning the development of genetic engineering had a rapid effect on the sector. In October 1990, when all of the EU nations adopted two European Council

Directives governing the "contained use" and "deliberate release" of genetically-modified organisms (GMOs) into the environment, this restrictive piece of legislation led to a shift in entrepreneurship. Between 1998 and 2004, no new GMOs were authorized for planting or for use in the EU until the issues of traceability and labeling were resolved. In April 2004, new European directives[1] allowed the trade of some approved genetically modified (GM) products, but the new rules from Brussels stated that all ingredients—above a threshold of 0.9 percent—that contained or consisted of genetically modified organisms, or that contained ingredients produced from GMOs, must be labeled and traceable. The rules also set up a centralized procedure to consider applications to grow and market GMOs in the European Union. According to a 2003 European Commission survey of private biotech companies and public research institutes (European Science and Technology Observatory 2003), 39 percent of the respondents canceled research projects on GMOs during the years 1999–2003, and in the private sector alone, 61 percent of respondents canceled research projects in this field. Between 1998 and 2001, the number of notifications for GMOs' field trials in the EU decreased by 76 percent.

The figures showed that foreign direct investment (FDI) was also affected by up to 50 percent between the first term of 2003 and the first term of 2004. This trend reached far beyond Europe, and the biotechnology sector was affected worldwide. For example, in 2003, Uganda refused to grow a disease-resistant GM banana out of fears it would lose its European market. The company Bayer CropScience India announced in November 2004 that it had stopped all its work on creating new GM crops in India and would instead concentrate on conventional plant breeding. It also stopped all trials of GM oilseed rape in Australia. In effect, entrepreneurship was changing to accommodate EU regulation. If we look at R&D as an indicator of entrepreneurship, we find that the decline was stronger in the U.S. than in Europe. Research from Ernst & Young's European Investment Monitor (Ernst & Young 2003) showed a decline of 11 percent in R&D in the European biotechnology industry between 2002 and 2003, and a decline of 17 percent in R&D in the US biotechnology industry for the same period.

In *The Competitive Advantage of Nations* (1990), Porter explained that it was by understanding the reasons for the ability of a nation's firms to create and sustain competitive advantage in particular industries that one could find clues to the competitive advantage of nations. In the ten years that elapsed since Porter's book was first published, globalization changed much of the facts and data, and the rise of the EU as an economic entity meant that the notion of the competitive advantage of nations had shifted toward that of blocs (China, the EU, the U.S., etc.). At the same time, entrepreneurship in emerging fields, and particularly in the biotech and nanotech sectors, had become one of the drivers of the competitive advantage of the economic blocs. The case of biotechnologies suggests that while the U.S. regulatory system initially fostered entrepreneurship by providing a favorable environment for entrepreneurs, the constraints of EU regulation led to shifts in entrepreneurship alongside wealth destruction (Auplat 2006). In other words, entrepreneurs who complied with the EU regulatory environment probably found themselves in a better position than those who did not in the long run.

What does this mean for the development of nanotechnologies? The empirical analysis of archival sources and patenting activity led to the conclusion in chapter 12 that although there were still important structural differences between the approaches of the three major players—the U.S., the EU, and Asia—there was a move toward an alignment of the major areas of nanotechnology development. Another finding was that sustainable development was increasingly perceived as a driver for nanotechnology entrepreneurship. In this chapter, the study of regulatory practices illustrated how they accompany and, to a certain extent, shape entrepreneurship. One provisional conclusion is therefore that institutional emergence—the creation of new institutions or the modification of existing ones—reflects the two distinctive features of nanotechnology entrepreneurship: the importance of sustainable development, and the progressive alignment of areas of entrepreneurship. Chapter 14 will probe a particular case of nanotechnology institutional emergence to further investigate this process.

Part IV

Innovative Practices and Nanotechnology Institutional Emergence

14 Introducing the Case of the ETC Group

I stated in the introduction that the objective of this book is to show that institutional emergence in nanotechnologies marks a rupture from previous forms of institutional emergence. This rupture is not produced by one group of stakeholders, one set of actions alone, but by a complex web of interactions emanating from five groups of stakeholders: entrepreneurs, commentators, scientists, policy makers, and end users. I postulated that the connecting element between all these interactions—the source of the rupture from previous forms of institutional emergence—is increasing awareness of the importance of sustainable development.

In the previous pages, I have looked successively at the actions of scientists, of entrepreneurs, and of policy makers. It is now time to start integrating the parameters that have emerged to analyze how they combine together. To do so, I look at one particular aspect of institutional entrepreneurship dynamics: the integration of civil society in the process of institutional emergence. I focus on one case study: the ETC Group (pronounced *et cetera*). The ETC Group—regarded as one of the first NGOs to have brought nanotechnology to public attention—was virtually unknown a decade ago. Through a longitudinal analysis conducted over the course of a five-year period, I examine its role and that of its leader in the early design of new regulatory frameworks and institutions around the development of nanotechnologies. I show that a combination of innovative practices, the use of an entrepreneurial toolbox, and bricolage allowed it to become an OPP (obligatory passage point), as developed by Callon and Latour (Akrich, Callon, and Latour 2006)) in the process of institutional emergence, thereby giving birth to a new form of institutional dynamics wherein decision making is no longer top-down as it used to be but is instead becoming a more complex, multistakeholder process.

A large body of literature has emerged since the 1990s to consider how new forms of organization arise and become established in the organizational community. The literature represents a very wide array of theoretical perspectives, and various streams of research have sought to explicate the nature of contexts and processes that may generate new organizational forms. This analysis situates itself in the theoretical perspective of an emergent social systems view that considers variations in organizational forms to be the products of embedded social-organizational interactions (Romanelli 1991). I consider the rise to prominence of the ETC Group as part of a social movement acting outside of institutional channels to promote change in society, what Gamson(1995) defined as a 'challenging group.' Considering that in the case of the ETC Group, success means being accepted by institutions as a legitimate organization addressing valid concerns, I build on Gamson's work to assess the different variables that may have contributed to the group's success in achieving its goal (Gamson 1990).

As mentioned previously, the development of nanotechnologies is recent, but it diffuses very fast and very broadly, and is likely to bring deep and radical changes to society. It constitutes an ideal empirical field for analyzing the emergence of new institutions in real time because it raises questions linked not only to entrepreneurship and innovation, but also to risk management and to ethics (Hunt, Mehta, and eds. 2008). Risk management may well be the cornerstone of institutional entrepreneurship in nanotechnologies. It may be *the* explanation for the interactions of the different groups of stakeholders that I mentioned in my model. The development of any new technology raises issues relating to risk management, i.e., to the process of measuring or assessing risk and developing strategies to manage it (Johnson and Covello 1987; Cothern 1995; Vogel 2003; Singh 2006). The difficulty with nanotechnologies is that it is at the present time extremely difficult to measure or assess the risks linked with the development of nanostructures. There are two major reasons for this. First, the majority of the nano effort is still at the research stage, which means that the risks that may or may not exist concern essentially the scientists themselves. They are only really beginning to develop nanostructures in a stable and measurable way, and toxicity assessment standards and protocols are largely nonexistent (Wiesner et al. 2006; AFSSET 2010; Johnson 2010). Second, because nanoparticles are so small, it is very difficult to trace them and to assess their inherent risks. At present, scientists lack the tools to trace and measure the potential risks induced by these particles. One speaks here of monitors as well as of modeling capacities. The only certainties are that nanoparticles are so small that they can cross most barriers and filters, including cell barriers within living organisms, and that due to quantum effects, it is still difficult to predict their behavior in many cases (Oberdorster 2004; Wick et al. 2010). There are an increasing number of safety-related studies concerning the use and development of nanotechnologies, but the results diverge hugely (European Commission 2009). As Roure and Dupuy, the two coauthors of a French report on a nanotechnology industrial prospective, put it:

> We stumble collectively against the lack of results of impact studies, against the absence of risks characterization, against the difficulty to assess their nature and financial valuation in terms of insurance liabilities and against the obvious shortage of adequate conceptualization. (2004, 39)

It was in this context that Pat Mooney and his NGO, the ETC Group, Action Group on Erosion, Technology and Concentration, came to prominence in the nanotechnology debate, with a twofold position on nanotechnology:

1. Requesting the establishment of a moratorium on the commercialization of new products containing novel, engineered nanoparticles until lab protocols are established to protect workers, and until regulations are in place to protect consumers and the environment
2. Requesting the creation of an International Convention for the Evaluation of New Technologies (ICENT), a new body dedicated to track, evaluate, and accept or reject new technologies and their products

The first public reference to nanotechnology coming from the ETC Group dated from 2001, with a report called *The ETC Century* (Mooney 2001). The NGO had started its activities a few years earlier by focusing on seeds, genetic erosion, and corporate concentration in the fields of pesticides, and on seeds and intellectual-property monopolies over life-forms and biotechnology issues. Pat Roy Mooney, the cofounder and executive director of the ETC Group, was born in Brandon, Manitoba (Canada), in 1947. What is common knowledge about him is that he is a father of five and is legally blind, and that he did not receive university training. He is also known to have lived most of his life on the Canadian prairies. Yet, to use the terminology of Callon (1986), the NGO and its executive director had become in 2006 an OPP in the nanotechnology debate. How did this come about, and what roles did the man and his organization play in this debate? What was their part in the emergence of new institutions surrounding the development of nanotechnologies? Could the answers to these questions provide an outline of a theory of a new form of institutional emergence centered around specific knowledge structures?

15 Theoretical Background

Institutional theory provides a theoretical model of analysis for the process of institutional construction that fits well with the development of nanotechnologies as a field (Lawrence, Hardy, and Phillips 2002; de-Holan, Phillips, and Lawrence 2004; Phillips, Lawrence, and Hardy 2004). However, the exploration of the impact of Pat Mooney's innovative practices on the emergence of new institutions leads us to revisit some of the theory's conclusions on organizational analysis, notably those concerning embeddedness and institutional isomorphism (DiMaggio and Powell 1983; DiMaggio 1988; Jepperson 1991). Indeed, the findings of this chapter will illustrate two important phenomena. First, they reveal that the extent of Pat Mooney's influence on the institutionalization of nanotechnology is directly linked to his innovative practices. Second, they corroborate those works showing that innovation processes themselves are more like bricolage than the traditional linear model of innovation would suggest (von Hippel 1977; Garud and Karnoe 2003; Baker and Nelson 2005). Indeed, in the ETC case, although the succession of events and decisions appeared to follow a clear strategy, it was also a combination of chance opportunities. The analysis also demonstrates that it was due to the 'unembedded' nature of its strategy that the NGO received sufficient support to emerge as an OPP (obligatory passage point) in the nanotechnology debate.

The notion of OPP was initiated by the work of Callon, Latour, et al. on actor network theory (ANT) and on the sociology of translation in the 1980s (Callon, Law, and Rip 1986; Akrich, Callon, and Latour 2006). This work focuses on the institutionalization of science and technology, without a particular emphasis on nanotechnologies. However, I consider that the model they developed is quite useful in our case. According to them, the institutionalization of science and technology rests on a chronological process that includes several steps. The time sequence of this process is as follows: first, there is the appearance of a series of apparently disconnected facts. These facts give rise to *a controversy* over their value and validity. Controversy is followed by *translation*, an operation that consists of creating a common language of reference so that the previously unrelated entities can be connected and seen in interaction. The process of translation gives unity and intelligibility to things that would otherwise have remained obscure and disconnected. It also allows the institutionalization process to move into the domain of public understanding. This process happens through the *network*, an ensemble of meta-organizations including human and nonhuman *actants* that interact together.

For Callon, Latour, et al.—and as they illustrate with various cases—it is not an intrinsic value that lies at the root of scientific institutional emergence, but the *settlement of a controversy* over the validity of new facts through the interactions of the various actants of a network. In other words, science and technology first emerge as disconnected facts and gradually stabilize and acquire ontology through the settlement of controversies concerning their validity. This analysis has led the researchers to propose a method in ten steps to account for institutional emergence or change. The initial step is the situational analysis of the context in order to identify all the actants. The final one concerns the spreading of the network and the necessity of transparency to keep the actants' trust. The turning point of the institutionalization process is what researchers call the OPP. Although Callon and Latour have not specifically explained it, they probably borrow the term 'obligatory passage point' (in French, *point de passage obligé*, or PPO) from the French maritime vocabulary, where it referred to a particular place that sailing boats had to go through if they wanted to reach a certain destination. For example, the Cape of Good Hope was an OPP for Portuguese sailors wanting to reach the Far East. Even if they knew it was a dangerous place, they could not ignore it, and they could not take another route. They had to go through it. In the ANT context, the obligatory passage point may refer either to a physical or virtual location, i.e., to a geographical place but also to a specific situation gathering certain actors, or to a type of discourse. It is the place (geographical, physical, or institutional) or the discourse that cannot be ignored or done without during the institutionalization process.

Callon built his theory from the empirical case study of the domestication of scallops in the bay of St Brieuc in Brittany (Callon 1986). Hoping to help local people develop a new business activity, three researchers had tried to farm scallops in Brittany. These three men were institutional entrepreneurs because they wanted to modify the institutional context in which they operated. Fishing was a real institution there, with its legal, social, environmental, and political dimensions, and they wanted to create a new one: scallop farming. The researchers were also an OPP because their discourse gave its orientation to the venture that they had initiated. They had become the 'heads' of several populations—experts, fishermen, and crustaceans—and all the stakeholders of the debate over scallop farming in France knew about their experience, referred to it, and acted according to it (whether to agree with it or to reject it).

Similarly, in 2006, the ETC Group and its executive director had become an OPP in the nanotechnology debate. They could be seen as institutional entrepreneurs because they were trying to modify the institutional context in which they operated, either with the creation of a new institution (the ICENT—International Convention for the Evaluation of New Technologies) or with the modification of existing ones (changing regulation to request the establishment of a nanotechnology moratorium). Because the other stakeholders could not ignore them, they became an OPP. Whether the stakeholders thought the ETC Group created a danger or not, whether they liked it or not, they could not do as if it did not exist.

In this chapter, I have built on some findings of the entrepreneurship literature to interpret the success of an institutional entrepreneurship venture. Regarding the tantalizing question of how to account for successful entrepreneurship, the entrepreneurship literature has produced interesting insights, which I put to the test of institutional entrepreneurship. Among the salient features is the importance of innovation—technological innovation as well as resource management innovation (von Hippel 1977; Aldrich and Zimmer 1986; Zucker and Darby 1998; Pinchot and Pellman 1999; Gaglio 2004). I find that this, in combination with a first-mover advantage (Lieberman and Montgomery 1988, 1998), constitutes the most valid explanation for the success of the institutional entrepreneurship venture of Pat Mooney.

16 Methodology

In this chapter, methodology rests on extensive archival research, complemented by in-depth interviews (Foddy 1993; Rubin and Rubin 1995; Denzin and Lincoln 2000; Wengraf 2001). Archival research included regulatory archives and media archives, as well as industrial archives and the archives of the ETC Group. In their discussion about structuration and institutionalization, Barley and Tolbert (1997) stressed that the investigation of how patterns of interactions lead to the emergence of a new institution was a difficult task and that to account for the links between actions and the emergence of new institutions, it was important to have a dynamic model rendering structuration through time. In my study, the interviews were conducted over a period of four years, between 2003 and 2007. I met Pat Mooney, the head of the ETC Group, several times, and combined the information coming from these interviews with interviews of key stakeholders in the nanotechnology debate: researchers, entrepreneurs, journalists, and nanotechnology officials of various national and intergovernmental institutions. The interviews took place in Belgium, Canada, France, the UK, and the U.S. (see appendix for a chronology of nanotechnology involvement by the ETC Group). A template for the analysis was drawn from DiMaggio's statement asserting that institutional changes "arise when organized actors with sufficient resources (the institutional entrepreneurs) see in them an opportunity to realize an interest that they value highly" (DiMaggio 1988, 14). So I began my study of the ETC Group as an institutional entrepreneur by identifying the actors and looking at how they were organized. Then I looked at what resources they used and how they mobilized them and, finally, I analyzed how they "realized an interest that they value highly."

According to a study released by the Centre for Applied Studies in International Negotiations, there were in 2006 about 60 nongovernmental organizations involved in nanotechnology advocacy globally (Lee and Rwanda 2006). The authors' classification was rather loose, involving trade unions and consumer associations, or regional chapters of the same NGOs. It also listed—without discrimination—organizations seeking to enhance the development of nanotechnologies as well as organizations aiming to restrict it. This list was helpful only to a certain extent, as it included entities that were either too dissimilar to draw any conclusion, or else too close together (e.g., regional chapters of the same NGO). In order to find answers to the question of why the ETC Group and its executive director had become an OPP in the nanotechnology debate in 2006, I needed to compare the NGO with other similar ones. Thus to make comparisons relevant, I focused on NGOs with a similar mission statement to that of the ETC Group, which was the following:

- Dedicated to the conservation and sustainable advancement of cultural and ecological diversity and human rights
- To support socially responsible developments of technologies useful to the poor and marginalized
- To address international governance issues and corporate power

Source: ETC Group, 2006.

As it so happened, for all of these NGOs, the focus on nanotechnology was only a *part* of their activities on a broader scale. None of them had been started in an ad hoc way simply to discuss the case of the development of nanotechnologies. Their common denominator was that one of their stated objectives was *to preserve the environment and encourage sustainable development.*

This pool produced a list of a dozen NGOs. To assess their relative import, I checked the number of Internet occurrences of their name in association with the term *nanotechnology* using a search engine. Although it was a rather crude way of estimating the importance of nongovernmental organizations in the nanotechnology debate, it clearly showed a group of five NGOs standing out—with a number of occurrences above approximately 200,000—then a sharp drop to fewer than 60,000 hits for all the others. The drop was so significant that it validated this rather simplistic approach (Figure 16.1).

Because my purpose was to understand the innovative practices that had enabled the ETC Group to reach prominence, I decided to consider only the first five NGOs in my study.

NAME	Number of occurrences
ETC Group	1,000,000
Friends of the Earth	991,000
Environmental Defense	829,000
Greenpeace	295,000
Sierra Club	185,000
OXFAM	63,000
WWF	47,400

Figure 16.1 NGOs involved in the nanotechnology debate. Compilation dated February 27, 2007.
Source: author

17 Circumstances and Practices

ACTORS

Figure 17.1 (on the following page) summarizes key data relating to the five NGOs. In line with Actor-Network-Theory (ANT) I consider them as actors

The first two features that stood out when I compared the ETC Group with the four other NGOs were its size and its management model. The ETC Group was extremely small in size. It had eight permanent staff who did not all work full time. Accountancy services were outsourced to a consultant. The board of trustees, which oversaw the organization, was composed of a president, a secretary-treasurer, and six other members (this number could vary a little). Also, the ETC Group was the only organization without members. Its management did not want it to grow and considered that if it were to grow, it would be by accident, not by intent. This raised the following questions: How could such a small structure have reached such prominence in the nanotechnology debate? What innovative processes enabled it to compensate for its lack of human capital? Was the small number itself part of the efficiency process? If so, how?

RESOURCES

In the discussion about resources, one common starting point was financial resources. Figures 17.2 and 17.3 compare the budgets of the five NGOs and the origins of their funding. One word of caution here: NGOs have greatly improved the transparency of their accounts, simply by having them audited and made public—recently, in some cases. As an example, it was only in August 2005 that Greenpeace International had 100 percent of its accounts audited (Greenpeace 2005). However, the arcane nature of NGO funding still makes it quite difficult to interpret. For the sake of clarity, Figure 17.3 only mentions the financial data relating to the *international* NGOs cited. But Greenpeace International, for instance, is an NGO with affiliates in over forty countries worldwide, each being run independently. The major source of funding for Greenpeace International comes from contributions

Name	Mission statement	Date of creation	Membership
ETC GROUP	Dedicated to the conservation and sustainable advancement of cultural and ecological diversity and human rights. To support socially responsible developments of technologies useful to the poor and marginalized. To address international governance issues and corporate power.	1984 in Canada under the name of RAFI. Took its current name in 2001	No members
FRIENDS OF THE EARTH	To challenge the current model of economic and corporate globalization To promote solutions that will help to create environmentally sustainable and socially just societies	1969 in the United States Became an international network in 1971	The world's largest grassroots environmental network Approximately 1.5 million members and supporters around the world
ENVIRONMENTAL DEFENSE	To link science, economics and law to create innovative, equitable and cost-effective solutions to society's most urgent environmental problems. To pay special attention to U.S. environmental problems and to America's role in causing and solving global problems, and to share approaches internationally.	1967 in the United States	Speaks for 500 000 members
GREENPEACE	To expose global environmental problems, and to force solutions for a green and peaceful future, using non-violent, creative confrontation. To ensure the ability of the earth to nurture life in all its diversity.	1969 in Vancouver, Canada Became the Greenpeace Foundation in 1972	An estimated 2.8 million financial supporters
SIERRA CLUB	To explore, enjoy and protect the wild places of the earth. To practice and promote the responsible use of the earth's ecosystems and resources. To educate and enlist humanity to protect and restore the quality of the natural and human environment. To use all lawful means to carry out these objectives.	1892 in the United States	America's oldest and largest grassroots environmental organization Over 750 000 members

Figure 17.1 NGO websites and annual reports (2007).
Source: Author's compilation.

from the regional offices. There is, however, a large gap between the budget of Greenpeace International and the total of the budgets of the regional offices. In 2005, the total net unrestricted assets of Greenpeace International amounted to € 45.7 million, but the total of all the unrestricted assets of the Greenpeace groups globally was €129.4 million. Similarly, when NGOs provide the sources of their funding, it is not easy to interpret the statements. For instance, some say they do not accept money coming from corporations or governments, yet they are subsidized by foundations that are themselves directly funded by corporations (Figures 17.2 and 17.3).

Until 2005, the ETC Group relied only on funding coming from governments or foundations (Figure 17.4). This meant that the very small staff did not need to spend time or resources on securing funding. This is itself an important element in explaining its efficiency in relation to its small size. The NGO also went against the usual idea (at least in the U.S.) that the best way to achieve efficient lobbying was to set up offices at the foot of Capitol Hill. Its headquarters were until 2004 in Winnipeg, a small town in Manitoba, Canada, quite remote from both political and scientific central spheres.

Figure 17.4 shows that approximately a third of the ETC Group's funding came from organizations supported by the Canadian government, another third from organizations supported by the Swedish government, and the rest from American private foundations. Clearly, the financial resources of the NGO were small. This was in itself a reason for being innovative in their management. Financial efficiency was a leitmotif of the NGO, and finding solutions for achieving the most cost-efficient clout was part of its DNA. But this also meant that the prominence achieved by the NGO could not be ascribed to its financial resources. So we have to look elsewhere to understand what brought the ETC Group to the forefront of the nanotechnology debate. We can also renew the questions asked after the description of the actors (Figures 17.1 and 17.3): How could a structure with so few financial means have reached such prominence in the nanotechnology debate? What innovative processes enabled it to compensate for its lack of capital?

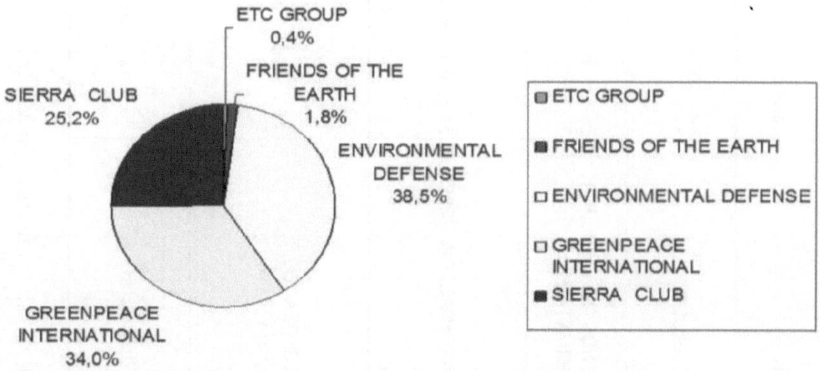

Figure 17.2 NGO websites and annual reports (2007).
Source: Author's compilation.

Name	Financing: total net unrestricted assets	Sources of finance	Place of registration of the international NGO	Headquarters
ETC GROUP	587,200 euros	100% coming from grants from various foundations.	the Netherlands	Ottawa, Canada Winnipeg, Canada, until 2004
FRIENDS OF THE EARTH INTERNATIONAL	2.38 million euros	90% coming from individual members and supporters	the Netherlands	Amsterdam, the Netherlands
ENVIRONMENTAL DEFENSE	51.7 million euros	59% coming from members, 32% coming from grants from foundations, 1% coming from government and other grants, the rest from bequests from members of the Osprey Society, investment income and other revenue	the US	New York, the US
GREENPEACE INTERNATIONAL	45.7 million euros	Income of Greeenpeace International: Over 95% coming from grants and donations from individuals or charitable foundations, the rest from merchandising and investment. Does not accept funding from governments or corporations.	the Netherlands	Amsterdam, the Netherlands
SIERRA CLUB	33.9 million euros.	Income of Sierra Club Foundation: Mostly coming from private funds and individual contributors. Accepts funding from corporations.	the US	San Fransico, CA the US

Figure 17.3 Sources of NGO funding.
Source: Author's compilation.

ETC Group Revenue in euros - 2005	
Swedbio	195 190
CS Fund	62 418
Canadian International Development Agency	96 853
IDRC	67 012
Lillian Goldman Charitable Trust	46 425
Ford Foundation	39 664
Educational Foundation of America	40 346
CBDC	17 137
Other	22 787
Total	**587 832**

American source	
Canadian source	
Swedish source	
Other	

Figure 17.4 ETC Group revenues in euros (2005).
Source: Author's compilation.

RESOURCE MANAGEMENT

This chapter is about an early-stage process of institutional creation. There was in 2006 no 'nanotechnology industry' or 'nanotechnology sector.' Regulation has often been described as a 'pillar of institutionalization' (Scott 1995), and there was at that time no specific regulation pertaining to nanotechnologies. Scientists even argued whether there was such a thing as 'nanotechnology research.' It is at this early stage, indeed, that one can observe the work of institutional entrepreneurs: Pat Mooney and his NGO took part in many of the debates, and their voice was heard. Between 2003 and 2006, 421 articles were written in the English language about the ETC Group and its executive director, in publications as varied as the *Taipei Times*, the *Washington Post*, *Forbes*, and the *St Petersburg Times*. Pat Mooney spoke about nanotechnologies at the European Parliament in Brussels, as well as at various hearings and testimonies worldwide. His name and that of his NGO appeared alongside those of top scientists at conference proceedings. It is impossible to say that they were responsible for a particular orientation in the debate—for example, that they contributed to the cancellation of some specific nanoprojects or to the introduction of particular pieces of legislation. Neither their demand for a moratorium on the commercialization of engineered nanoparticles or their request for the creation of an International Convention for the Evaluation of New Technologies (ICENT) was successful. Yet they achieved, in spite of their minuscule size and lack of financial means, the position of OPP (obligatory passage point) in the debates: they participated in the major discussions, and references to the position of civil society invariably included allusions to them. In the following section, I analyze how their innovative practices allowed them to make up for their lack of human and financial capital.

RHETORIC

Institutional researchers have begun to explore the importance of rhetoric and discourse in the institutional process (Phillips, Lawrence, and Hardy 2004; Suddaby and Greenwood 2005; Maguire and Hardy 2006; Phillips and Malhotra 2008). In Pat Mooney's case, rhetoric was a tool that enabled the ETC Group to aggregate a particular community, which in turns provided access to various institutional levies (science, policy, regulation, etc.). As was mentioned previously, Pat Mooney, the NGO leader, had no university training. He acknowledged that his research director, Hope Shand, played an instrumental role in putting together the reports that contributed greatly to the recognition of the NGO, thanks to her technical writing and research skills. One thing he did not say was that he was himself extremely articulate, with a particular gift for arriving at catchy formulas. The rhetoric of the ETC Group had some very distinguishable features.

First, it introduced neologisms. The ETC Group regularly introduced new terms or concepts. Not all endured—far from it—but a few did, and they contributed to strengthening the identity of the NGO. Two neologisms that have remained were *biopiracy* (introduced in 1993), and *terminator seeds* (introduced in 1998). Both terms were coined when the main concern of the NGO was the loss of agricultural genetic resources and the erosion of biodiversity, resulting in a desire to protect small farmers against the overriding force of big agribusinesses. Pat Mooney explained that in the order of things, they identified the technology, gave it a name, and then went to obtain support to ban it. In its latest reports, the ETC Group continued to introduce neologisms. Some examples include *atomtechnology, BANGS* (bits, atoms, neurons, and genes), and ICENT (International Convention for the Evaluation of New Technologies). Only time will tell whether these will remain in the lexicon, but the NGO leader stated clearly that coming up with good names was an important part of the success of 'projects,' as he called them.

Second, it combined detailed—mostly accurate—scientific facts with anecdotes that tended to be unexpected. Several scientists pointed out the somewhat bizarre rhetoric of the NGO, where "Ninety percent is real material and then they came up with something that is disconnected" (Stuart 2003). This technique may irritate scientists, but it attracted the attention of the general public. It contained a good deal of puns, which drew the reader's notice and kept him or her entertained, or at least alert. The language itself was a mixture of very topical and sometimes colloquial references with articulate notions requiring some academic culture spelled out in a rather elaborate style. A few examples will illustrate this point; for instance, a report on the increasing concentration of the world's resources in the hands of a few began with the heading *the Equity Erosion E=TC2* and went on to explain:

> It does not require an Einstein to recognize the new equation of power: The exponential erosion of our biosphere—coupled with the erosion of our ability to understand the biosphere—coincides with a similarly exponential expansion of our technical ability to manipulate large living systems, safely or otherwise. What remains of diversity, and the technologies thrust upon diversity, are falling into the hands of corporate oligopolies. (Mooney 2001, 18)

A report on attempts to alter the earth's climate stated that

> Paul Crutzen returned to the debate, stirring up a real tempest in a teapot in August 2006 when he wrote an 'editorial essay' in *Climatic Change* magazine calling for active research into the use of 'sub-micrometer'-sized sulfate-based aerosols to reflect sunlight in the stratosphere in order to cool the earth. Crutzen, a professor at the

Figure 17.5 Some examples of ETC cartoons.
Open sources 2007:
http://www.etcgroup.org/en/materials/cartoons.html;
www.shtig.net; http://www.drooker.com/; reypage(at)shaw.ca.

> Max-Planck-Institute for Chemistry in Mainz, Germany, opines that high-altitude balloons and artillery cannons could be used to blast sulfur dioxide into the stratosphere, in effect, simulating a volcanic eruption. (Mooney 2007, 6)

These examples illustrate the combination of plays on words, disconnection between the seriousness of the subject and the tone used, and rather sophisticated technical references.

Third, the ETC Group rhetoric involved regular historical comparisons and references. These were carefully chosen to appeal to the largest possible audience, spanning regional cultural differences. They tickled the reader's brain and called on his or her general culture, creating what could be called a 'spirit of connivance': the reader's understanding of the reference rested on a shared cultural heritage, and the rhetoric therefore created a sentiment of belonging—belonging to an international cultural elite. They also gave the demonstrations a 'commonsense' feel. Put into a historical perspective, the NGO statements acquired a timeless value. The following is an example of a very common rhetorical practice:

> "The main reason for productivity gains [during the Renaissance] was technological progress. . . . ," historian Carlo Cipolla insists, looking back on the explosion of wealth during that period. The productivity of Italian weavers doubled and then tripled—even without the textile machinery that became the trademark of Britain's Industrial Revolution centuries later. Gutenberg's first printers churned out three hundred pages a day. By the end of the Renaissance, a printer could produce four times that amount. Between 1350 and 1550, English iron production rose seven- or eightfold. Many of the Renaissance advances came in the areas of shipping and trade. Before Columbus, the crew-to-cargo ratio was one sailor for every five or six tons. The Dutch achieved a ratio of one man per ten tons by the end of the Renaissance. Five hundred fifty years later: info, nano, and bio. (Mooney 2003, 7)

Fourth, the ETC Group used cartoons and drawings in its reports and publications (see Figure 17.5). They followed the same pattern as the word rhetoric, calling on the reader's general culture and relying on a mixture of humor and the clash between technical accuracy and incongruity.

This analysis complements the work of researchers such as Suddaby and Greenwood (2005) and Maguire and Hardy (2006), who show the importance of rhetorical strategies in aggregating a community around a common theme—an essential step toward institutional change. Indeed, it was by rallying a large enough number of people who showed interest in the issues relating to nanotechnology that the ETC Group could begin the process of bringing the topic to public tables of discussion.

INTERNET TECHNOLOGY

Internet technology was at the core of both the innovative practices and the management of the NGO. Not surprisingly, the ETC Group had close relationships with Sweden, one of the pioneering and most advanced nations in terms of Internet use. This gave it an edge, and the NGO relied on Internet Technology from the start. In terms of management, Pat Mooney explained that the three major assets of Internet Technology use for the group were that the technology was very cheap compared to the usual means of communication, that it enabled it to overcome time differences easily and to stay in touch permanently and, finally, that it allowed it to act quickly, one of the critical factors of success according to the NGO leader. The early use of Internet Technology was also part of the NGO's innovative practices. It enabled the ETC Group to spread its work quickly and broadly. Contrary to many organizations that sell their reports, from the beginning, the NGO made everything it produced freely available on the Internet. This gave it an advantage, as it was a precursor in doing so, and it also strengthened the initial recognition of their work. Finally, Internet Technology was from the start a cheap and easy source of information for the group. The NGO was innovative in that it understood this sooner than others.

LANGUAGE OUTREACH

The NGO may have been a precursor in its early widespread use of IT, but its outreach would not have been the same if it had not been coupled with the vision of communicating in Spanish and Portuguese as well as in English. It had started doing so when it was focusing on seeds, and the credibility it had acquired with this issue gave it a strong advantage at the beginning of the nano issue (Hardon, Montecinos, and Roberts 2005). From the start, the ETC Group communicated in English and Spanish, and constant efforts were made to further increase the number of languages in which publications or statements were released, in French, Portuguese, Swedish, and German, among other languages. The initial will to reach both the Spanish- and English-speaking communities globally had an enormous influence on the visibility of the NGO. At the beginning of the Internet Technology revolution, most of the data circulating on the Internet was restricted to English-speaking communities, so data coming from the English-speaking world that was made accessible in Spanish had great appeal to Spanish-speaking people, particularly in South America. As a result, Pat Mooney and his team were able to gain global recognition—and in a way some sympathy—quickly. In 2005, out of the ten countries that most frequently accessed the ETC website, five were Spanish speaking.

NETWORKS

The literature on entrepreneurship is extremely clear that social ties are of vital importance for the success of new business ventures (Aldrich and Zimmer 1986; Dubini and Aldrich 1991; Uzzi 1997; Hoang and Antoncic 2003; Yli-Renko, Autio, and Sapienza 2001; Clarysse et al. 2005). This study shows that social ties also constitute an element of paramount importance for explaining *institutional* entrepreneurship. It confirms previous findings of the entrepreneurship literature on the necessity of a combination of weak and strong ties to guarantee both innovation and stability. These are key features of successful business ventures (Lawrence, Morse, and Fowler 2005; Karra, Tracey, and Phillips 2006). They are also essential to the success of *institutional* entrepreneurship ventures. The ETC Group was in fact run by a very small group of people who had known each other for thirty years, had evolved together, and had followed the same ideals, although not always the same paths. The core of the group consisted of three people: Pat Mooney, Cary Fowler, and Hope Shand, who began working together in the mid-1970s, at that time on seeds issues and the preservation of agricultural biodiversity.

In 2007, the ETC Group's staff was based in four locations: three people in Canada (Ottawa), two in the U.S. (North Carolina), two in Mexico (Mexico City), and one in the UK (Oxford). Pat Mooney and his staff considered that the group's small size was an important way of achieving efficiency because the chains of communication were simple and because personnel problems could be solved very easily, or prevented from happening altogether. Decision making could be very swift. To optimize this advantage, the NGO made the best use of Internet Technology to remain in constant contact in different locations and time zones.

One explanation for the incredible efficiency of the NGO in relation to its small structure and lack of human capital comes from what Latour et al. called 'actants' to denote the human and nonhuman actors giving shape to a network through their interactions (Latour, Woolgar, and Salk 1986). In the case of the ETC Group a key actant was the Dag Hammarskjöld Foundation. The stories of the NGO and of the foundation were closely intertwined. The Swede Dag Hammarskjöld was the second Secretary-General of the United Nations, from 1953 to 1961. The foundation that bears his name was set up in his memory in 1962 in Uppsala, Sweden.

The ETC Group was built on the same model as this organization in terms of human resources: the Dag Hammarskjöld Foundation had an extremely small staff of only two professionals and three secretaries, one of them part time. It also had a very long-lasting structure with very little turnover: the current director, Olle Nordberg, took over from his predecessor Sven Hamrell in 1994. Sven Hamrell had been the foundation's director since 1967 and had been working with Olle Nordberg since 1969. The beginning of the collaboration between the Dag Hammarskjöld Foundation and the ETC Group dates back to 1981. Sven Hamrell was the first president of RAFI, the ancestor of the ETC Group, and Olle Nordberg had been on the ETC board since its creation. Both organizations had the same model of management, and both had the same strategic vision: identifying causes and fighting for them. They both very much dealt in foresight. They also both aimed at altering institutions.

The Dag Hammarskjöld Foundation constituted a key actant in the emergence of a network of interactions that might fashion the nanotechnology debate because it provided the ETC Group with resources. The resources were not strictly financial ones. The Dag Hammarskjöld Foundation was not on the list of the funding agencies of the NGO. However, it had been instrumental in publishing the first reports of the ETC Group and cohosting conferences with it. According to a study by the California Institute of Public Affairs, the success of the Dag Hammarskjöld Foundation itself, in relation to its extremely small size and budget, rested a great deal on "mystery, loyalty and democracy" and on "having a lot of friends and drinking together" (Trzyna 1990). This operating mode could also possibly explain the success of the ETC Group.

These very strong ties, involving a close-knit community, were what held the NGO together, what gave it its stability. Other strong ties existed with the funding agencies, as well as with the governments of Canada and Sweden. Apart from these, the other ties were actually rather weak. For example,

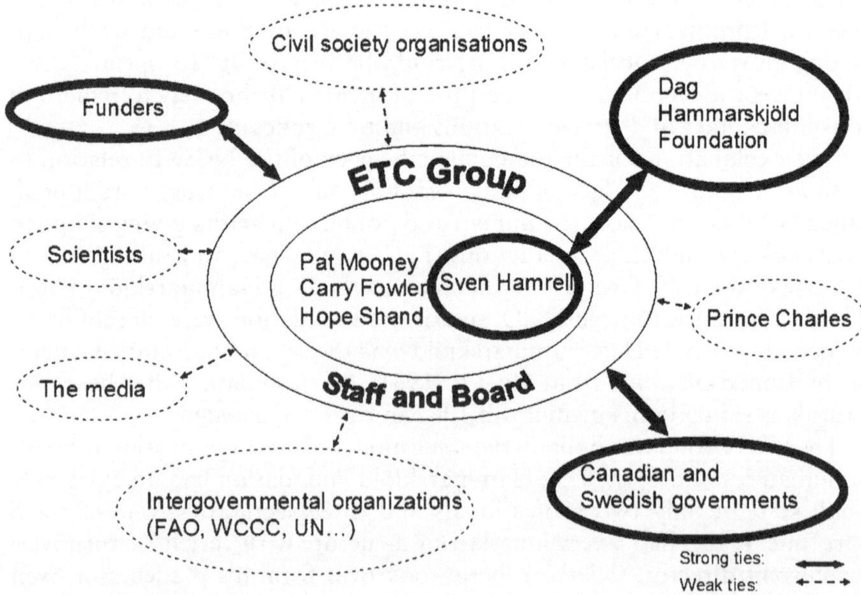

Figure 17.6 Summary of the network of interactions around the ETC Group. *Source:* author.

although the ETC Group worked closely with the NGO Environmental Defense on some issues, it did not hesitate to lambaste it for its proposed collaboration with DuPont on nanotechnology in early 2007. These weak ties connected the ETC Group to the fringes of other organizations, and corroborated the findings of institutional researchers that DiMaggio and Powell's analysis of embeddedness had to be somewhat qualified: neither isomorphism nor embeddedness (DiMaggio and Powell 1983) could give a full account of institutional change, and it was by looking at the fringes of organizations that one could find the origins of institutional change (Greenwood and Suddaby 2006).

The ETC Group was at the center of a network of interactions that was progressively lending ontological coherence to the institutionalization process. Figure 17.6 outlines the main ties that constituted this network of interactions.

In this chart, embeddedness is illustrated with the ties to the Dag Hammarskjöld Foundation, funders, possibly governmental ties, and those with civil-society organizations. One distinctive feature of the ETC Group was that most of its offices were shared with other, larger NGOs. This was an efficient way of reducing costs, of course, but that wasn't the only reason. The close proximity of the other NGOs boosted the knowledge capacity of the ETC Group, but more importantly, it also provided a multiplier effect that increased diffusion and impact.

Fringe relations covered all the other ties. Because the discourse of the ETC Group was new to the world of Prince Charles, the heir to the throne of the United Kingdom and other Commonwealth realms, he was intrigued enough to invite some of its members to tea to debate nanotechnology issues with representatives of the Royal Society. Although the ETC Group staff members were clearly not scientists—or *because* they were not scientists—eminent researchers such as Professor Vyvyan Howard, who was until 2004 president of the Royal Microscopical Society, agreed to coauthor a report with them (ETC Group 2003). The same can be said of its ties with the UN and other intergovernmental agencies. Pat Mooney made it clear in his interviews that he wanted to stay on the periphery of these organizations—that he did not seek to be too closely connected to any of them. Combining weak and strong ties, embeddedness, and fringe relations was part of his strategy.

CHANCE, FIRST MOVER ADVANTAGE, AND LIFE-CYCLE ISSUES

I have already discussed that the success of Pat Mooney's venture can be explained in part by the theory of the first-mover advantage (Lieberman and Montgomery 1988; Porter 1998). Indeed, the ETC Group was the first NGO to voice public concern about the risks associated with nanotechnologies. In an interview he gave me in May 2004, the NGO leader elaborated on the NGO's move to the issue of nanotechnologies. He explained that his group had first spotted the issue more or less by chance in the 1980s as part of its routine scouring of the scientific press. It had found the issue fascinating and started building a thin file on it. At that time, the hot issue was biotechnologies, so the NGO's major activity was to rally other NGOs to the cause of biotechnologies. It had looked at biotechnologies long before the other NGOs had and spent years trying to tell other NGOs that they should take this seriously and look at the implications of these new technologies. It found that by the late 1990s it had been successful in the sense that a wide range of NGOs had taken on the issue of biotech. As a very small organization, it was then faced with the choice of staying involved in that issue or getting involved in a new cause, which was what it did. Although retroactively people stated that there was a connection between nanotech and biotech, this was not the way things had happened concerning the involvement of the ETC Group with nanotechnologies.

This analysis provides empirical support to Romanelli's analysis of emergent social systems (Romanelli 1991). It corroborates the theory that technological innovation is a dynamic social process that, as it unfolds, creates the resource space that will support the new firms reflecting new organizational forms. It also supports the theory of life cycles. Organizations compete for resources and appropriability—an economics term which refers to the ability to capture profits through innovation. For NGOs it means having an edge in identifying new causes and rallying public support for them. In all cases, organizations have to move on to new targets once they have reached the peak of their efficiency at their own level (Noymer 2002).

18 Findings

Assessing the relative success of the institutional entrepreneurship venture of Pat Mooney and his NGO was no easy task, especially as the process was not complete when the field research ended in 2007. At that time, there had not been a moratorium on nanotechnology research, and there had not been a creation of an ICENT (International Convention for the Evaluation of New Technologies). However, what the NGO had achieved was stirring up a controversy and being part of the debates about the creation of new institutions, mainly regulatory institutions, but also ethics groups overseeing technological developments. What it could *claim* to have achieved was the following:

- Contributing to putting nanotechnology regulation on the agenda
- Contributing to slowing down research in some areas
- Contributing to a new societal approach of technological progress

Interestingly—and this was definitely part of the NGO's strategy—the ETC Group was very modest about its role. It did not boast about its participation in influential meetings, and the lists of key events relating to the various milestones of the development of an institutional framework around the development of nanotechnologies tended to overlook its own involvement (see, for example, ETC Group 2005). Yet it can be said with certainty that Pat Mooney could be considered an institutional entrepreneur: he played an active part in the emergence of an institutional framework around the development of nanotechnologies. What is of particular interest is that he apparently had everything going against him in this venture: He was a person with a handicap (functionally blind), with no university training, no particular scientific background, living in a remote place in Canada, and with quantitatively very few human or financial assets. It was by using the entrepreneurial toolbox of business entrepreneurs that he achieved what he did. The important elements of his success were his

character and personal dynamism;
capacity to motivate;
strong ties and networks;

vision; and
use of IT.

Using these tools—and language outreach was part of the innovation—in an innovative way, the ETC Group aggregated a community of supporters who provided access to decision makers. Most observers agree that one of the main differences between the early development of nanotechnologies and that of biotechnologies that took part thirty years earlier was that civil society was involved in the discussions from a much earlier stage with, therefore, a different institutionalization process. The ETC Group was the first NGO to bring civil society into nanotechnology discussions, and one of its roles was to give some legitimacy and recognition to the voice of NGOs in institutionalization processes.

Where does this take us in relation to the model of nanotechnology institutional emergence presented in Figure 2.2? The term *commentator* on Figure 2.2 was meant to have a very generic meaning. A commentator can have several roles. It may be someone who observes and comments on something, but also someone who reports and analyzes events. The ETC Group belonged to this group of stakeholders because of its self-described role: The NGO's mission statement set out to address international governance issues and corporate powers in order to support socially responsible developments of technologies useful to the poor and the marginalized, as well as to devote itself to the conservation and sustainable advancement of cultural and ecological diversity and human rights.

The advocacy role of the NGO consisted of commenting on the development of nanotechnologies and reporting its own analysis of the situation. Several conclusions can be drawn here. I explained earlier that in order to capture the specifics of institutional emergence around the development of nanotechnologies, I had chosen to develop a few case studies illustrating relevant features, considering that the ability to replicate and generalize my findings would be the test of the validity of the method. I postulated in the introduction that institutional emergence in nanotechnologies marked a rupture from previous forms of institutional emergence. The integration of civil society in the process is one sign of this rupture. In this chapter, I offer some explanations for understanding how this rupture could come about. One set of explanations related to the means used by NGOs to become part of the institutionalization process: they acted like business entrepreneurs. In other words, NGOs become institutional entrepreneurs by using the same toolboxes as entrepreneurs. Although the term *business plan* never came up in discussions, it was obvious that the ETC Group followed clear internal guidelines with specific 'customer' (audience) targets, 'product/services' (what Pat Mooney called projects), identification and launch timelines, resource mobilization and risks, and exit strategies. As business entrepreneurs do, Pat Mooney identified and exploited opportunities for new ventures, and he combined existing resources and used new ones in order to do so.

One major difference between Mooney and a business entrepreneur was that his venture was clearly not about making money. Generally speaking, business entrepreneurs explicitly seek to make a profit—this is their unequivocal motivation. In the case of institutional entrepreneurs, it is more difficult, because their common motivation is not so straightforward. Put simply, they aim to create new institutions or to alter existing ones, but there is no single obvious reason for doing so. DiMaggio referred to "an interest that they value highly," but without providing details. Is it power? Is it social recognition? Or is it something else? When the director of the Dag Hammarskjöld Foundation was asked how the foundation decided what to do next, how it identified the fields on which it focused, he replied that it basically worked on hunches and intuition (Trzyna 1990). Asked the same question, Pat Mooney observed that he and his team tended to go where their loves took them. Although these replies may not tell the whole truth, they do confirm the findings of Garud and Karnoe that there is always a certain amount of bricolage in path creation (Garud and Karnoe 2001). They also lead to another set of explanations to account for a rupture in institutional dynamics. This time, the findings do not relate to the *means* allowing institutional entrepreneurs to shape the process of nanotechnology institutional emergence. They shed light on the reason *why* this can happen.

I explained in the introduction that nanotechnologies had arrived at a point in time when technological progress and social evolution had reached a point of convergence with increasing awareness of the importance of sustainable development for the economy as well as for the general public. This global awakening to the importance of sustainable development, coupled with the widespread use of information technology, modified the model of institutional emergence that Callon et al. described with previous new technologies. As noted at the beginning of this chapter, the common denominator of all the NGOs involved in nanotechnology advocacy in 2006 was that one of their stated objectives was to preserve the environment and encourage sustainable development. So it may be inferred that the common "interest which they valued highly" was precisely the preservation of the environment and encouragement of sustainable development. This was exactly what the general public was beginning to feel the need for. What happened between 2003 and 2006 was that an NGO such as the ETC Group was the most visible structured group available to relay the public's concern for sustainable development in the field of nanotechnologies. It could comment on some of its aspects. So, to summarize, one of the characteristics of institutional emergence in nanotechnologies was that for the first time the voice of civil society groups was heard from the start. This could happen for two reasons: the cause of sustainable development had acquired enough legitimacy (even if it was very little) to be impossible to ignore, and there was a structured organization available and altogether acceptable by all parties to voice concerns.

Part V

New Institutionalization Processes

Risk Management and Public Engagement

19 A Chronology of Nanotechnology Institutional Emergence

In this chapter, I pursue my exploration of the coindustrial and institutional emergence of nanotechnology by focusing on end users. The March 2011 update of the inventory of nanobased consumer products compiled by the Pew Charitable Trusts and the Woodrow Wilson International Center for Scholars in the PEN project contained 1,317 products or product lines. The inventory had grown by over 500 percent (from 212 products) since its first release in March 2006. However, nanoproducts were still used by only a *very small* proportion of the global population at that time. So when using the term 'end users,' I am giving it a very broad meaning. It refers not only to people who actually used nanoparticles, but also to potential users. It also includes people who might be affected by nanoproducts not by choice. As was explained earlier, nanotechnologies could in theory affect absolutely everything, as they correspond to modifications of matter at the atomic level. So they could potentially diffuse everywhere and concern the whole of mankind under various forms, including matter that was released incidentally in the environment, or by-products of it. 'End users' in this sense refers to all those who would be affected willy-nilly by nanotechnologies as they began disseminating in our daily lives.

As the process of diffusion was spreading, numerous groups of stakeholders became involved with them. These stakeholder groups had their respective value systems for characterizing the importance of various parameters. For example, start-up companies focused on the benefits of growing markets for their nanobased products. Insurance companies needed to know how much to charge for premiums based on the risk they were taking on for insuring the nanotechnology industry. Users of products containing nanomaterials—notably in the health sector—wanted to enjoy the benefits of the new products, but were also concerned about the effects of exposure. What all these stakeholders had in common was their involvement with end users. However, their concerns were mostly very diverse, and it was the interactions of their different constituencies—each managing nanotechnology-related risks with their varying value systems and priorities—that shaped the institutionalization of the field.

In order to assess the role of end users in the process of institutional emergence, I focused on the case of public-engagement experiments. Public-engagement experiments probably constituted the most legible indicator of the early interactions of end users with nanotechnologies. They also reflected their concerns and their perceptions of risk issues. They corresponded to a new form of participatory risk management that included a mechanism of aggregation of the expectations of the various stakeholders in the risk-management process, and this constituted a rupture from previous forms of risk management. It also characterized the development of nanotechnologies. Therefore, it confirmed the rupture in institutional emergence postulated in the introduction.

Once again, this chapter is firmly anchored in institutional theory. It integrates the findings of institutional economics that established relations between the performance of economies and institutions and institutional change (North 1990). Institutional theory also provided the theoretical model of analysis for the process that I examined, a process of institutional construction around the development of nanotechnologies (de-Holan, Phillips, and Lawrence 2004; Lawrence, Hardy, and Phillips 2002; Lawrence and Suddaby 2006; Lounsbury 2007; Phillips and Brown 1993; Phillips, Lawrence, and Hardy 2004; Tracey 2005). My objective was to contribute my own interpretation of the institutionalization process by focusing on risk-management dynamics. After showing how risk was gradually institutionally integrated in the development of nanotechnologies, I analyzed some of the characteristics of the new form of risk management that was emerging. I found that this new form of risk management marked a rupture from previous forms of risk management and deducted that the development of nanotechnologies signaled a new form of institutional emergence.

In his landmark study of institutions and organizations, Scott (1995), aiming to organize a hitherto somewhat nebulous field of research, devised a classification scheme that could encompass the numerous existing conceptions of institutions. He set up a typology around three types of analytical elements that could be found in various institutional carriers: regulative, normative, and cognitive. He called these institutional 'pillars,' meaning that they were at the heart of institutions. Much discussion has taken place about the validity of this typology, and Scott himself considered that the three pillars were on the whole interdependent and mutually reinforcing. Yet even though some critics later argued that some elements might be placed in more than one section and that the borders between the three pillars were hazy, by considering these independently, Scott was able to shed light on the essence of institutions in a way that had not been done before (Dacin 1997). In this study, I follow Scott's route in that I look at sets of analytical elements that are found in the various institutional processes accompanying the emergence of nanotechnologies as a field. But I depart from his work in that my aim is not to categorize various forms of institutions. Instead, I seek to unpack a particular process of institutional emergence. So my typology is not about the sets of elements that constitute institutions. It is about sets of elements that are at the heart of the institutionalization process. I call these pillars 'pillars of the institutionalization of nanotechnology.'

Chronologically, it seems that it was North America and the UK that produced the first bases of two pillars of the institutionalization of nanotechnology as I define them. These were an institutional pillar-base and a nongovernmental pillar-base.

INSTITUTIONAL PILLAR-BASE, PHASE 1

In 2001, the U.S. created a single program centralizing the major research efforts relating to nanotechnology, the NNI (National Nanotechnology Initiative). Before this decision, nanotechnology R&D had been a haphazard smattering of various unconnected initiatives, in the U.S. and everywhere else. The NNI pursued the following aims:

- Maintain a world-class research and development program aimed at realizing the full potential of nanotechnology.
- Facilitate transfer of new technologies into products for economic growth, jobs, and other public benefit.
- Develop educational resources, a skilled workforce, and the supporting infrastructure and tools to advance nanotechnology.
- Support responsible development of nanotechnology.
  Source: NNI.

The report published in May 2005 by the U.S. President's Council of Advisors on Science and Technology (PCAST 2005) found that the U.S. was the world leader in nanotechnology as well as the acknowledged leader in nanotechnology research and development. It concluded that the NNI appeared well positioned to maintain United States leadership going forward thanks to its coordinated interagency approach to planning and implementing a federal R&D program, and it supported the development of the initiative.

The creation of the NNI was followed by the publication of the S. 189 Act, which defined the program in section 2 as

> (10) ensuring that ethical, legal, environmental, and other appropriate societal concerns, including the potential use of nanotechnology in enhancing human intelligence and in developing artificial intelligence which exceeds human capacity, are considered during the development of nanotechnology by
>
> (A) establishing a research program to identify ethical, legal, environmental, and other appropriate societal concerns related to nanotechnology, and ensuring that the results of such research are widely disseminated;

(B) requiring that interdisciplinary nanotechnology research centers established under paragraph (4) include activities that address societal, ethical, and environmental concerns;

(C) insofar as possible, integrating research on societal, ethical, and environmental concerns with nanotechnology research and development, and ensuring that advances in nanotechnology bring about improvements in quality of life for all Americans; and

(D) providing, through the National Nanotechnology Coordination Office established in section 3, for public input and outreach to be integrated into the program by the convening of regular and ongoing public discussions, through mechanisms such as citizens' panels, consensus conferences, and educational events, as appropriate. (U.S. Congress 2003)

With the creation of the NNI and the following S. 189 Act, the U.S. set a double trend: the need to coordinate all nanotechnology research, and the necessity to assess and address societal implications throughout the development of nanotechnologies. This was the first basis of the institutional pillar of the institutionalization of nanotechnologies. In the wake of the American launch, most nations and regional powers such as the EU recognized the importance of coordinating nanotechnology research and assigning it some guidelines. Nowhere did the word *risk* appear in these U.S. texts institutionalizing nanotechnology as a coordinated national enterprise. Yet, the first texts already mentioned the necessity of engaging the public in the development of the new technology.

NONGOVERNMENTAL PILLAR-BASE, PHASE 1

Just as these important institutional steps were taken, on the nongovernmental level in 2002, the Canadian-based NGO Action-Group on Erosion, Technology and Concentration asked for a moratorium on the commercialization of new products containing novel, engineered nanoparticles until lab protocols were established to protect workers, and until regulations were in place to protect consumers and the environment (ETC Group 2002). This could be seen as the source of the nongovernmental pillar-base. The ETC Group was indeed the first NGO to draw attention to potential problems linked with the development of nanotechnologies. Interestingly, the word *risk* did not appear once in this first text either.

INSTITUTIONAL PILLAR-BASE, PHASE 2

After North America, the UK government commissioned in 2003 the Royal Society and the Royal Academy of Engineering to study potential applications of nanotechnology and their possible environmental, health and safety, social, and ethical implications. The report was published in 2004. It referred to the call for a moratorium by the ETC Group and included questions about some risks posed by manufactured nanoparticles, along with arguments on the need to review chemical regulations to ensure that they encompassed nanotechnologies.

> Under current UK chemical regulation (NONS) and its proposed replacement under negotiation at European level (REACH), the production of an existing substance in nanoparticulate form does not trigger additional testing. We have recommended that this regulatory gap be addressed by treating nanoparticulates as new substances, thus requiring additional testing, under both NONS and REACH. (Dowling 2004, 82)

In the report's final chapter, "Recommendations," the word *risk* appeared four times, and in chapter 8—the chapter dealing with regulatory issues—the word appeared twenty-nine times.

The report did not find any justification for imposing a ban on the production of nanoparticles. However, as a precautionary measure it recommended that releases to the environment be minimized until effects were better understood. It also mentioned the importance of public and stakeholder dialogue. It recommended that the UK government initiate a properly funded public dialogue around the development of nanotechnologies at a stage when such discussions could inform key decisions about their development and before deeply entrenched or polarized positions appeared.

NONGOVERNMENTAL PILLAR-BASE, PHASE 2

In 2003, the NGO the ETC Group published a new report in collaboration with a renowned toxicologist, Dr Vyvyan Howard (ETC Group 2003). The report included a literature search relating to the effects of nanosized particles on human health and the routes by which nanoparticles could enter the body, and mentioned the word *risk* fourteen times. Another report published earlier that year, *The Big Down*, showed thirty mentions of the term (Mooney 2003).

This chronology of events extends previous research on the increasing importance of risk management in society (Beck 1992; Bennett and Calman 1999) in that it highlights how the awareness of its importance has become *part* of institutional emergence. This in itself marks a rupture from other technological developments, where risk management issues appeared *after* the development of new technologies, as was illustrated by cases as varied as nuclear research (Campbell, Einhorn, and Reiss 2004), DDT (Maguire 2004). and asbestos (White 2004). It also reveals that civil society gained admission in the process of institutional emergence. First, official structures such as the Royal Society acknowledged the input of nongovernmental organizations. Second, public structures explicitly emphasized the need to engage the public *from the earliest stages* in the development of the technology. Chapter 20 is the study of this process, which will be detailed further in the form of an analysis of nanorelated public engagement experiments.

20 Risk Management and Institutional Emergence

The development of any new technology raises issues relating to the process of measuring or assessing risk and developing strategies to manage it (Cothern 1995; Singh 2006; Vogel 2003; Johnson and Covello 1987). From even the earliest times, humans have likely been aware of the benefits of group action in reducing risk. In forming a social structure, individuals who helped others to recover from hardships (such as a fire destroying their dwelling) understood that if needed, a similar benefit would be extended to them by other community members. Approaches to formal risk mitigation have become increasingly sophisticated over time. However, there are numerous examples of early approaches to transferring or distributing risk. The practice by Chinese merchants of distributing their wares across many boats to limit the loss due to any single vessel's capsizing when traversing treacherous rivers dates back as far as the third millennium BC.

Alongside these basic approaches to sharing risk emerged the notion of risk assessment, as people realized three things. First, the 'value' of goods or persons exposed to dangers was not identical; second, some perils were more dangerous than others; and third, some hazards were more likely to happen than others. Not surprisingly, the evolution of risk assessment closely paralleled the evolution of an insurance sector. Historical examples include the establishment of such insurance organizations as Lloyds in the seventeenth century to insure cargo during sea transportation, or Nicolas Barbon's insurance company to insure buildings against fire in the aftermath of the Great Fire of London in 1666 (Hanson 2002). Risk is now generally defined as being a function of both the expected losses that can be caused by an event and of the probability of this event. The harsher the loss and the more likely the event, the greater the overall risk. These two factors, loss and the probability of the loss, are typically treated as sequential, independent events so that, in simplified terms, quantification of risk is expressed as:

$$\text{Risk} = (\text{probability of an accident}) \times (\text{losses per accident})$$

Along with risk assessment goes the notion of risk management (Louisot and Gaultier-Gaillard 2007; Anderson et al. 2009). Risk research has developed as society has become increasingly aware of its capacity—or incapacity—to manage risk, particularly when new potential sources of risk are presented. One characteristic of risk management is that risk is always acknowledged and assessed *a posteriori*. Scientific proof can only be put forward when harm has already been done. This is the only way to obtain an estimation of an unobserved quantity on the basis of empirical data. It was only after the London fire destroyed 13,200 houses and left some 70,000 Londoners homeless that Nicolas Barbon's idea to insure buildings against fire could first germinate and then become successful (Wainwright 1953). It was only in 2007, when Hurricane Dean approached oil-production facilities off the coast of Mexico, that the oil market began focusing on the potential for weather-related supply disruptions (Tan and Evans 2007). The paradox is that risk can only be envisaged in the light of past events and yet bears on the probability of the occurrence of future events. This paradox leads to three types of risk management.

MANAGING THE KNOWN KNOWNS

This type of risk management is usually well mastered. For example, the dangers of fire are known. It is common knowledge that fires and their associated losses depend upon two distinct factors—the likelihood that a fire hazard exists and will produce a fire, and the exposure of a given value (represented by the people, property, heritage, environment, or disruption affected) to the fire. So risk management will be about reducing the likelihood of starting a fire and increasing fire protection.

MANAGING THE KNOWN UNKNOWNS

What is known about the H1N1 virus is that during the 1918–20 Spanish flu epidemic, it caused anywhere between fifty and one hundred million deaths globally, and killed about a third of the European population of the time (Patterson and Pyle 1991). What is also known is that the 2009 swine flu outbreak was due to a new strain of subtype H1N1. Experts agreed that another pandemic was inevitable and possibly imminent (WHO 2007), as influenza pandemics can be expected to occur three or four times each century, when new virus subtypes emerge and are readily transmitted from person to person. What is unknown is when the pandemic will strike, and how deadly it will be. Here, risk management is about preparing oneself as best one can. This includes among other things trying to develop vaccines, finding operational solutions to keep vital business activities going if populations are decimated, developing logistics connected with mass-fatality management, and preventing panics. The main challenge of managing the known unknowns is keeping public trust (Irwin, Jones, and Stilgoe 2006).

MANAGING THE UNKNOWN UNKNOWNS

By definition, it seems impossible to manage unknown unknowns, as no one can have any insight into unknown risks linked with unknown hazards. Unfortunately, the development of nanotechnologies falls under this category. Incomplete information regarding nanomaterial properties and their potential impacts makes it particularly difficult to measure or assess the risks linked with their development. The concept of risk associated with nanotechnologies was first put forward by Eric Drexler in 1986 when he described a "grey goo" scenario in which self-replicating "nanobots" might consume all the carbon on the planet while building more of themselves (Drexler 1986). The feasibility of such a scenario has been debated at length and remains contentious. However, the more immediate question of whether or not nanomaterials might be possible contaminants posing risk to human health and environment was raised publicly for the first time in December 2000.

> Let's envisage a release in the environment of these molecules, because of an accident in the production plant, or simply at the end of the life cycle of the manufactured product. Will they become a health hazard like asbestos? What will be their impact on ecosystems? (Wiesner 2000)

Increasing concern over the possible risks presented by nanomaterials led to some of the first experiments in public engagement on the topic during the period 2002–4. This particular form of institutional emergence was explained by various factors:

- There was a genuine desire of governments to "do things right" from the start (European Commission and Nanoforum 2004; European Union 2003; U.S. Congress 2003).
- Previous academic studies had shown the importance of involving the public in policy making in the face of technological uncertainty (Auplat 2006; Callon, Lascoumes, and Barthe 2001; Fiorino 1990; Godard et al. 2002).
- The industry was very wary about repeating the same mistakes that had hampered the development of biotechnologies (Auplat 2006; Chataway, Tait, and Wield 2006; Gaskell, Ten Eyck, and Jackson 2004) and was cautious about the possibility of creating legacy issues that might return to haunt corporations, as had occurred in the case of hazardous-waste legislation and litigation in the U.S.
- Similarly, the insurance sector was very cautious and expressed concern at covering products that might turn into asbestos-like problems (Swiss Re and Hett 2004).
- Policy makers were told that people are more and more inclined to be risk averse if they are not in control of the risk or if they do not trust the people who manage risk (Currall et al. 2006; Irwin, Jones, and Stilgoe 2006).

21 A Case Study of Public Engagement

The study of the relationship between risk management and public-engagement experiments in the nanotechnology debate rested on a comprehensive analysis of existing archival sources. Two reports—one emanating from French sources and one from British sources—attempted to establish a panorama of initiatives concerning public engagement undertaken between 2003 and 2007 in the nanotechnology field. The data presented concerned essentially European and American experiments, as there did not seem to have been many other instances of consequence elsewhere in the period concerned. I synthesized the various features of each type of experiment, checking my findings with focused interviews with key stakeholders of the public-engagement experiments. Finally, I drew conclusions about new dynamics in risk management. These conclusions were twofold: they bore on the concerns of end users with the development of nanotechnologies, and they applied to the impact of new risk-management dynamics on institutional emergence.

In 2005, the UK government published its Outline Program for Public Engagement on Nanotechnologies (OPPEN) as a response to the Royal Society and Royal Academy of Engineering report (Dowling 2004). The Dowling report had concluded that nanotechnology offered tremendous potential, but that more research was needed to discover any negative, as well as positive, effects of nanotechnology. It also asked for regulation to minimize any future risks. One of the objectives of the OPPEN program was to fund a study of public-engagement initiatives; NEG (the Nanotechnology Engagement Group) was established for this purpose.

NEG planned to document what was learned from a series of attempts to involve members of the public in discussions about the development and governance of nanotechnologies, and it did so in its report published in 2007, *Democratic Technologies?* (Gavelin, Wilson, and Doubleday 2007).

In France, the Department of Higher Education and Research and the Department of Industry commissioned in 2006 a report gathering the recommendations of the main nanotechnology stakeholders in order to get a picture of their positioning and recommendations toward nanotechnology (Cité des sciences et de l'industrie 2007). The report—entitled *Cahiers d'acteurs*—consisted of fourteen chapters, each one an independent report issued by one group of stakeholders. The stated objective was to gather the analyses of "the main stakeholders: industry, science and politicians." Interestingly, no mention was made of the opinion of citizens. However, half of the reports involved public-engagement experiments. One of the major differences between the *Cahiers d'acteurs* and *Democratic Technologies* reports was that the former was a gathering of primary sources, as each of the fourteen chapters came directly from the stakeholders themselves, whereas the latter spoke through the voice of the NEG's authors, as it was a digest of the various experiments they studied.

According to the NEG's report, there had been until 2007 some twenty public-engagement initiatives on nanotechnologies internationally. The report may not have accessed all available information, but no mention was made of any initiative in Asia or the Middle East. Out of twenty-three projects identified involving public engagement, fourteen came from EU countries, six from North American countries, and three from Australia and New Zealand. 'Public engagement' was defined as the different ways that

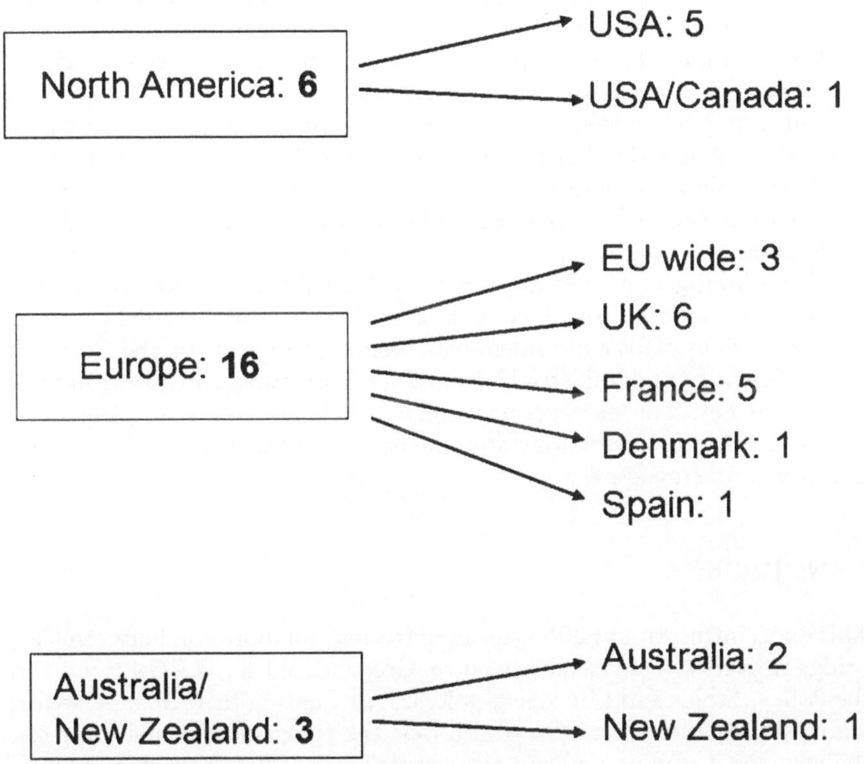

Figure 21.1 Countries involved in public engagement experiments: 2003–7.
Sources: NEG and Cité des Sciences et de l'industrie. Author's chart.

institutions interacted with the general public outside of formal democratic structures such as elections. The NEG study focused exclusively on forms of public engagement that involved dialogue between members of the public and scientists or decision makers, and excluded purely one-way forms of public engagement such as information campaigns or public-opinion polls. The scope of the projects varied a great deal, from a one-day workshop to a five-year project, and they spread out in time from 2003 to 2006.

Sticking to NEG's definition of public engagement, out of the fourteen initiatives documented in the *Cahiers d'acteurs*, seven focused on programs involving public engagement. There were some overlaps between the two reports' studies, and I identified a total of twenty-five different experiments meeting NEG's definition of public engagement, from which I drew my empirical data and established a typology of the major public-engagement initiatives in relation to nanotechnology from 2003 to 2007 (Figure 21.1).

GAMES AND SCENARIO PLAYING

The first public-engagement project—in 2003—was in the form of a game called *Democs-nanotechnology*. It was commissioned by the sixth EU Framework Programme (FP6) and the Wellcome Trust. Participants were dealt a series of hands of cards to read, and were asked to pick the ones that they felt were most important to the discussion. At the end of the game, participants stated their preferred policy position on a subject by choosing from four predeveloped policy positions or developing one of their own. The games usually involved teams of six to nine people (see www.neweconomics.org).

In another experiment in Australia, participants were divided into small working groups and given a hypothetical scenario kit to stimulate discussions about the social, economic, and environmental implications of nanotechnology. This experiment was supported by CSIRO, the Commonwealth Scientific and Industrial Research Organization, Australia's national science agency. The participants selected included community members, CSIRO staff, nanotechnology specialists, and government representatives (see www.minerals.csiro.au).

NANOJURIES

This experiment, run in 2005, resulted from a collaboration between Cambridge University Nanoscience Centre, Greenpeace UK, *The Guardian*, and the Policy, Ethics and Life Sciences Research Centre (PEALS) at Newcastle University. Modeled on legal trial juries, the experiment brought together the jurors, a group of sixteen participants from a West Yorkshire borough selected from the electoral register and via suggestions from youth and community workers.

Before turning to nanotechnology, the jury addressed a topic of their choice over eight evening sessions of two and a half hours each. They decided to look at young people, exclusion and crime in the local community. They then discussed nanotechnologies in a further ten sessions of two and half hours each. Six witnesses selected by an oversight panel provided evidence and in the last few sessions the jurors wrote recommendations for the future development of nanotechnologies in the UK. The recommendations were presented to an audience of policy-makers, scientists, journalists and social researchers at an event in London in September 2005.

DELIBERATIVE PUBLIC DEBATES

The format was designed for experts and specialists invited through organizers' networks to come and give evidence on various aspects of nanotechnology and then open discussions with the floor. The audience came freely, and most people attending had an interest or were involved in nanotechnologies. In the case of the French experiments, every meeting lasted about two and a half hours, and a report was released after each debate. A series of recommendations followed the debates, addressed specifically to policy makers, industry leaders, and academics (see www.vivagora.org).

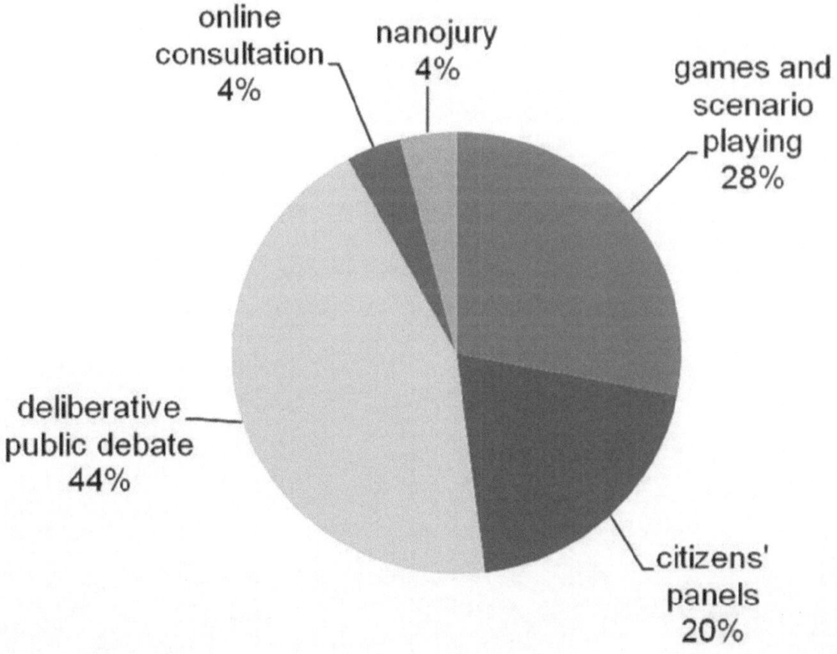

Figure 21.2 Formats of public engagement.
Source: author.

ONLINE CONSULTATIONS AND INTERACTIVE ASSESSMENT TOOLS

This methodology for public engagement was more and more routinely used in conjunction with each of the other types of experiments. Websites provided information about nanotechnologies, as well as interactive, more or less constructed spaces for chats or debates. However, in 2007, these practices were still too new to have received sufficient feedback to draw valid conclusions (See Figure 21.2).

22 Interpretation and Findings

In the vast majority of cases, the public knew nothing about nanotechnologies before their involvement in the experiments. Only 4 percent of these experiments concerned stakeholders who had a direct involvement in nanotechnologies, and then, the involvement was scientific rather than industrial or governmental. In most cases, participants engaged themselves on a voluntary basis, although NEG documented experiments where people were paid to participate or were enticed to do so by other means (for instance, to talk about other subjects that deeply concerned them).

An important aspect of the experiments was whether participants' thinking was directed or not. In the scenario playing, games options, nanojuries, and citizens' panels, those participating had to choose solutions from various options that were presented to them. This meant that their contribution was in fact in the form of a reaction to ideas that were submitted to them. Conversely, in the deliberative public debates and, to a certain extent, online consultations and interactive assessment tools, their participation was allowed to be more open. This distinction between open and closed discussion was a significant one. Indeed, history had shown that some of the societal issues accompanying previous technological innovations had not been anticipated by the scenarios of scientists or the early developers of these technologies. For instance, in the case of biotechnologies, it was farmers' associations that first mentioned some of the risks associated with gene patenting, such as the loss of cultural heritage for some nations, or the issue of ownership of living organisms. It was also nonspecialists who drew attention to the risks arising from the cohabitation between genetically modified crops and nongenetically modified ones. So it was important to note that nearly half of the public-engagement experiments left room for open discussion.

Experiments were organized by various sources. The vast majority (72 percent) originated from institutional sources, mainly from government or from science and research bodies. The others came from civil society, usually local community groups or groups with environmental preoccupations, or partnerships between industry and local community groups or NGOs. A word must be said about an experiment of a unique nature, that of *Cahiers d'acteurs N°13*. This report came from a gathering of trade unions representing workers involved in nanotechnology-related activities. In this particular case, the stakeholders did have knowledge of nanotechnologies, as well as a particular involvement with the technology.

Given the various differences just outlined—whether in the format of the public-engagement experiments, or in their location, their duration, the origins of the participants, etc.—the outcomes were altogether rather homogeneous. The same types of preoccupations, although not phrased identically, emerged. The outcomes of these various public-engagement experiments could be summarized as follows:

- On the whole, nanotechnology was perceived as a highly interesting area of innovation with a great potential for the improvement of society.
- Risk management was seen as one of the key issues of the development of nanotechnologies.
- There was consistent demand to complement the development of nanotechnologies with societal assessments and public engagement.

Risk was a very transversal topic in the experiments. However, the concept of risk itself was interpreted differently by the various participants (see Figure 22.1).

Most saw risks *in relation to the usage of nanomaterials*. Three subcategories emerged: health and especially human-health risks; risks of a degradation of the environment; and finally, risks of loss of freedom stemming from the potential for the miniaturization of monitoring devices. For most, assessing the risk/benefit ratio, as well as *whom* risks and benefits were most likely to affect, was the chief objective of risk management. They were concerned that nanotechnologies might bring more benefits than risks to more-developed countries but more risks than benefits to less-developed countries. Participants stressed the necessity to make sure that risks/benefits were distributed equally among more and less developed countries. Also, they wanted to take the notion of time into consideration: they did not want the development of technologies that might bring more benefits than risks in the short term but more risks than benefits in the long term. Although in some cases located in the UK, the debate was introduced particularly in the form of 'how to improve *British* society,' in all the experiments the answers were invariably of global purport.

The second understanding of risks was *in relation to policy making*. Many groups vented concern that, for various reasons, the industry, politicians, and regulators might not demonstrate enough transparency in the processes leading to decision making. The major concerns were twofold:

- How to reconcile industry's understandable need to preserve a measure of secrecy on some processes in order to maintain a

competitive advantage and public demand for more information and transparency.
- How to make sure that policy makers and regulators really understand and control the various aspects of the development of nanotechnologies.

These preoccupations did not correspond to clear-cut groups of stakeholders and spanned various groups. Similarly, calls for more regulation or for a moratorium on nanoresearch did not correspond to any particular group. When calls for more regulation were made, they covered the following areas:

- Chemical classification and labeling of nanoparticles as well as regulation to monitor nanoparticles releases in the environment
- Mechanisms of transparency in regulation and policy making
- Safeguarding of personal liberties

The fact that these public-engagement experiments took place, coupled with the development of the Internet and its multiplier effect on the shaping of public opinion, revealed how different risk management had become from what it had been only ten years before. This is where one might speak

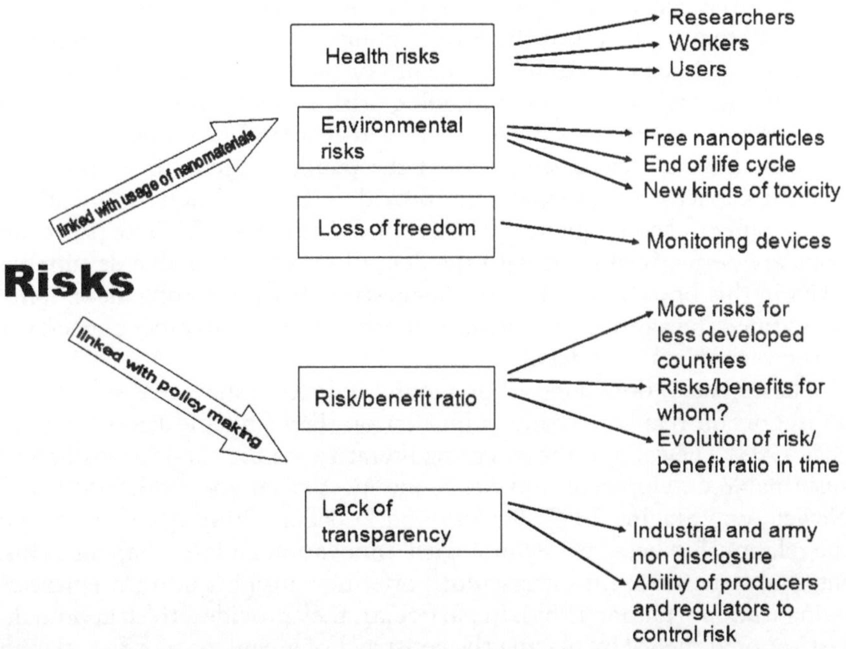

Figure 22.1 Summary of nanotechnology risk analysis based on the various public-engagement experiments studied.
Source: author.

of a reshaping of technological democracy. Just as democracy refers to the concept of 'rule by the governed,' technological democracy means that the stakeholders of a new technology are empowered to design its governance. In the case of the development of nanotechnology, the very fact that these public experiments took place with the support of existing institutions demonstrates that regulators now move toward a more inclusive method of policy making, which gives a new significance to technological and societal parameters. A major finding of the study carried out in chapter 21 was that a traditional top-down approach of risk management alone was no longer considered sufficient in addressing the concerns raised by emerging nanotechnologies (Irwin, Jones, and Stilgoe 2006). But this also meant that, by placing the construct of a new form of risk management at the core of institutional emergence, nanotechnology institutional emergence marked a rupture from previous forms of institutional emergence.

The case study of nanotechnology public engagement experiments showed that in the area of risk management, the development of nanotechnologies appeared to be generating a mechanism for aggregating the expectations of various stakeholders. The analysis of public-engagement experiments revealed that what the end users really formed a consensus on was the need to weigh the risks and benefits of new technologies for society before bringing them to market. Moreover, end users wanted to make sure that the benefits really outweighed the risks for all in the long run. It appeared that what was at the root of public concern was not so much scientifically defined risks to human health or the environment, but rather the need to know whether a new technology risk/benefit ratio would be clearly in favor of long-term benefits for society in general. The public-engagement experiments also revealed that what the public expected from regulation was the capacity to anticipate and provide a frame of action to deal with future potential consequences (the unknown unknowns). These preoccupations are perfectly aligned with the definitions of sustainable development given in this book's introduction, suggesting that sustainable development constituted the aggregating element of otherwise very distinct expectations of the various stakeholders.

These observations support previous studies that show that society plays an important regulatory role in innovation (Freeman and Perez 1990; Fischer 1993) They add to the emerging literature on the connections between sustainable development and the economy (Bansal and Roth 2000; Jaffe, Newell, and Stavins 2005; Potoski and Prakash 2004). Also, by studying the relationship between technological innovation and the shaping of institutional structures, the observations offer new insights into the emergence of institutions (Latour 1988). In particular, they provide a fresh approach of institutional theory by placing the construct of a new form of risk management at the core of institutional emergence (Phillips, Lawrence, and Hardy 2004; Romanelli 1991; Zucker 1977). By assessing the relative importance of the characteristics of nanotechnologies in the design of an institutional framework for their development—particularly in the areas of risk assessment and management—the study developed in chapter 21 sheds new light on the links between institutions and technology (Wiesner et al. 2006; Wiesner and Bottero 2007) and between the amount of organizational structure and performance in dynamic environments (Davis, Eisenhardt, and Bingham 2009). But most importantly, they support the theory that the rise of nanotechnologies marks the arrival of a new form of technological democracy, as will be developed in the next pages.

Part VI

Towards A New Form of Technological Democracy

23 Sustainable Development and New Governance Options for New Technologies

The purpose of this book is to provide a conceptual framework of the combined actions of various groups of stakeholders to show how nanotechnology institutional emergence marked a rupture from previous forms of institutional emergence. Rather than focusing on a particular group of stakeholders—the institutional entrepreneurial "heroes"—the perspective chosen aims to provide evidence that institutional entrepreneurship is a complex process that resulted from a multifaceted web of interactions. It reveals that with the development of nanotechnologies, the process of institutional emergence was coevolutionary with that of industrial emergence, and that there was a constant interplay of reciprocal influence between the two. The PEN venture (Project on Emerging Nanotechnologies) offered an illustration of this phenomenon. Its inventory of nanoproducts reaching the market showed that products falling under the heading 'health and fitness' were by far the most numerous and constituted more than half of the nanoproducts in the inventory. In the absence of specific nanotechnology regulation, and in the context of all the uncertainties mentioned at various stages, the development and marketing of new products tended to favor those that were the most likely to be accepted by the general public because they were products intended to improve one's health or to increase one's well-being.

Several specific issues were raised during the previous pages, and the time has come to look again at the initial questions that guided my work to see what answers were found.

WHAT MAKES THE NANOTECHNOLOGY CONTEXT A PROPITIOUS ONE FOR A RUPTURE FROM PREVIOUS FORMS OF INSTITUTIONAL EMERGENCE?

First, nanotechnology is a diffuse, cross sector technology that can potentially affect the whole of society, both horizontally (geographically speaking) and vertically (from the poorest to the richest, from the youngest to the oldest).

Second, it arrives at a point in time in the history of mankind when Internet technology and globalization mean that concerns are more and

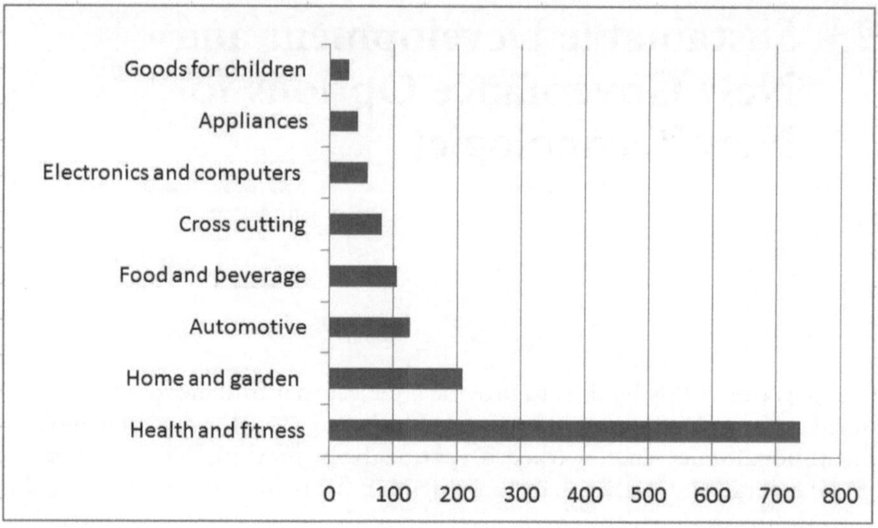

Figure 23.1 Number of nanoproducts on the market by category in 2011. PEN's data. *Source:* author.

more widely shared and that attempts to find solutions to emerging problems are more and more global.

Third, its distinctive characteristics render it both extremely attractive because of its potential to improve society and at the same time very worrisome because of the current inability to correctly assess the risks associated with its development. This paradox calls for new types of institutions that can cope with this duality.

WHAT ARE THE CHARACTERISTICS OF THE RUPTURE IN INSTITUTIONAL EMERGENCE? WHAT DO THEY REVEAL?

First, there is an overall global alignment of R&D priorities. This alignment is accentuated by the fact that economic studies on the promises of the technology are carried out by researchers or consulting institutions that tend to analyze data on a global scale and to render global conclusions. For example, the methodology described by Lux Research Inc. in its analysis of the nano value chain involved interviews with more than a hundred executives, researchers, and academics working to commercialize nanotechnology globally, as well as the integration of macroeconomic data from organizations such as the World Bank (Lux Research Inc. 2004). Alignment of R&D priorities contributes to the shaping of the field, and because R&D is closely related to institutional emergence (see chapter 4), institutional emergence also becomes a global affair. For example, since

FP6, the European Framework Programmes have clearly encouraged the emergence of European research and of a European Research Area (ERA) (Commission of the European Communities 2000). This marks a rupture from previous histories of technology emergence in which national interests were more significant.

Second, existing institutions put more weight on the shaping of the process than ever before. As one of the interviewees representing the European Commission stated, nanotechnology was the first case in history of a new technology getting such generalized and financially important institutional support for its development.

Third, the precautionary principle is underlying the regulatory framework of nanotechnology development. This time again, it is simply a question of historical conjunction: the precautionary principle came to underlie all European regulation in 2000, with *Communication from the Commission on the precautionary principle,* COM(2000) 1 (Commission of the European Communities 2000). As this text serves as a basis for all EU regulation, it covered the development of nanotechnologies from the start. This marks a difference with biotechnologies, for instance, because these had started developing *before* the adoption COM(2000) 1. Although COM(2000) 1 only applies at the European level, globalization and alignment of R&D priorities suggest that its effect is likely felt globally.

Fourth, civil society is integrated in the institutionalization process. As was demonstrated at various points in the previous pages, civil society was part of policy making during all the initial phases of the development of nanotechnologies. This happened at two different levels. One was when existing institutions gave space to civil society in the institutionalization process (see, for example, all of the official initiatives to facilitate the involvement of the public at various stages and in various circles). The other one was when civil society empowered itself to participate in the debates (see, for example, the case study of the ETC Group).

HOW DO THE VARIOUS STAKEHOLDERS ACTING AS INSTITUTIONAL ENTREPRENEURS COLLECTIVELY SHAPE THE INSTITUTIONAL EMERGENCE OF NANOTECHNOLOGIES?

The short answer is that nanotechnology institutional emergence corresponds to a new form of technological democracy. As a reminder, the meaning I give to technological democracy is that it is a type of governance whereby the stakeholders of a new technology are empowered to take part in the design of its governance. Just as democracy refers to the concept of 'rule by the governed,' technological democracy means that the governance of a new technology is designed by the combined actions of its stakeholders. It would be naïve to think that all is perfect in the world of democracies, and that democratic governance is simply the direct outcome of the work of elected bodies representing the people, where all citizens have an equal say in the decisions that affect their lives. Things are usually more complicated than that. However, democracy represents at our current stage of human evolution the best we can do to empower the governed to decide how they are ruled. Likewise, it would be naïve to consider that when new technologies emerge, all is transparent and all the stakeholders of these new technologies have an equal say in their use or development.

My point is that with nanotechnologies, there is an evolution that marks a rupture from previous cases of institutional emergence around the development of new technologies, and that this corresponds to a reshaping of technological democracy. What are the signs of this process?

Indeed, whereas previous cases of technological emergence tended to be driven entirely by developed countries, developing countries are participating in the nanotechnology process at various levels. One is by hosting various nanotechnology conferences. The following are a few examples from the events sections of the websites of Nanoforum (http://www.nanoforum.org) and Nanowerk (http://www.nanowerk.com), two leading nanotechnology portals and sources of information: the Fourth Bangalore Nano Conference was planned in India in December 2011 (see http://www.bangalorenano.in/, accessed March 2011); the Second Iran-India Joint Conference on Nanotechnology (IIJCN) was held in Iran in 2009 (see http://iijcn.ir/, accessed March 2011); the Third Biannual South African Nanotechnology Initiative (SANi) Conference, "NanoAfrica 2009," was held in February 2009 in South Africa (http://www.csir.co.za/enews/2009_feb/msm_05.html, accessed December 2010); the combined Nano-Aqua and Nano-Agri Conference, focusing on water and agricultural challenges for nanotechnologies, took place in Cairo, Egypt, in 2010 and 2012 (see http://www.sabrycorp.com/conf/nanoagri/10/overview.cfm, accessed April 2011).

By hosting these conferences—by giving a voice to the research of their own representatives—these different nations modify the process of institutional emergence: they provide a visible sign that it is being globalized. It is too soon to speculate about the actual changes that this new development will bring about, but it cannot be without effects. Research on tournament rituals, which shows how a highly charged ceremonial form may attract collective attention to a field and serve as a medium to create legitimacy, may be helpful in analyzing its impact further (Anand and Jones 2008; Anand and Watson 2004). What is certain is that the globalization of the organization of international conferences and the active participation of emerging countries have been a hallmark of the development of nanotechnologies since the organization of the first comprehensive conference on the topic in 1989. This was the first Foresight Conference on Molecular Nanotechnology. It was held in October 1989 in the U.S., was chaired by Eric Drexler, and was cosponsored by the Foresight Institute and the Global Business Network. Even at its inauguration, the conference drew participants from three continents and many disciplines.

We have seen that the development of nanotechnologies is giving birth to a new form of institutional dynamics wherein decision making is no longer top down as it used to be but is becoming a more complex, multistakeholder process. The process is still evolving, and I will develop in the next paragraphs one of its aspects: the loss of ownership of regulatory frameworks.

As the development of nanotechnologies is gathering speed, so too is the awareness of the importance of managing the risks potentially associated with nanoproducts gaining widespread notice—risks for consumers, for the environment, for the industry, or for insurers, to name just a few sectors (Wiesner and Bottero 2007; Singh 2006; Lloyd's 2007; Johnson 2010). The phrase 'nanotechnology risks' covers three different forms of risks: risks coming from nanoparticles, risks coming from public perception, and risks coming from regulation (Swiss Re and Hett 2004; Raje 2007). These risks are of very different types and affect different groups of constituents, as is made obvious if we ask the simple question: risks for whom?

Risks coming from nanoparticles mostly affect those who are exposed to them. This may mean human beings, animals, or the environment. In this case, risk assessment is based on a well-known model. It is a function of hazard and exposure. To assess risks with some accuracy, one faces the challenge of first knowing what is dangerous about a certain nanoproduct. This is the hazard. Can certain nanoparticles trigger cancers or other diseases? Can they harm certain ecological systems? As the technology is so radically new, there are no past experiences or conventional tools for measuring with certainty how dangerous given nanobased products are (European Commission 2009; Wiesner et al. 2006). The second variable is exposure. Exposure covers several areas. One may refer to researchers manipulating nanoparticles and, more generally, the exposure of workers producing nanoparticles. Beyond this, exposure relates to end users. When you use products containing nanoparticles, you are obviously exposed to them. For example, the European Commission estimated in 2006 that approximately 5 percent of cosmetic products contained nanoparticles. In August 2009, PEN listed ninety-eight consumer products containing specific nanoparticles in the food and beverage industry—for instance, food supplements or various products used to preserve food. Finally, exposure refers to the presence of nanoparticles in the environment, whether they have been released intentionally or not. Then, people may be exposed to them without wanting to or even without being aware that they are exposed to them. In all these cases, the major obstacle to overcome is once again the absence of data. Industrial producers are very reluctant to produce figures of actual production, and it is extremely difficult to assess with relative certainty the amounts of nanoparticles that are currently on the market in one form or another, which makes exposure assessment extremely uncertain.

Another form of risk associated with the development of nanotechnologies is risk coming from public perception (Anderson et al. 2009). In spite of a great deal of research done on risk analysis and risk-management theory

(Chavas 2004; Alexander 2008; Hopkin 2010), risks coming from public perception have not been investigated fully, and no satisfactory formulas have been found to account for them in a reliable way. Thus far, researchers use the case of the development of biotechnologies as a kind of benchmark for evaluating nanorisks coming from public perception (Gaskell, Ten Eyck, and Jackson 2004; Einsiedel and Goldenberg 2004; David and Thompson 2008). However, there are vast differences between the developments of the two forms of technologies, which constitutes a limitation to this approach.

The last form of nanotechnology risks is risks coming from regulation, and this is what is of interest to us here. The regulatory framework of nanotechnologies is volatile. At the moment, there is no certainty about the future of nanotechnology regulation (Bosso 2010). This uncertainty induces risks for nanotechnology entrepreneurs and for the emerging nanotechnology industry altogether. Indeed, the case of the development of biotechnologies revealed that regulatory changes entailed shifts in entrepreneurial activity accompanied by wealth destruction (Auplat 2006). If regulatory frameworks change, what was allowed at one point in time may not be so any longer, which means that investments may become useless if what they had been designed for cannot be developed. I build upon these findings to analyze the emergence of regulatory frameworks around the development of nanotechnologies. I show that the current regulatory procrastination that is accompanying the development of nanotechnologies is met with the emergence of a new form of regulation whereby voluntary norms such as international standards are taking precedence over national mandatory frameworks and are gradually acquiring new ontology, thereby leading to a loss of ownership of regulatory frameworks.

24 Governance and Norm

There are currently a large number of initiatives concerning *potential* normative frameworks to cover the development of nanotechnologies. Laas and Weil (2008) explored the nuances of each initiative. However, very little has actually been *enacted* in terms of regulation.

Now might be the time to say a few words on the REACH framework (2006/1907/EC) (European Union 2007), although this European law does not apply specifically to nanotechnologies. The term REACH stands for Registration, Evaluation, Authorization and Restriction of Chemical substances. This comprehensive regulatory framework was introduced by the European Commission in 2006 to replace some forty existing legal acts and create a single system for all chemical substances in Europe. Its main objective was to improve upon the protections in place for human health and the environment through the better and earlier identification of the intrinsic properties of chemical substances; this included a plan to find substitutes for the most dangerous chemicals currently in use. The law entered into force on June 1, 2007. After the cutoff date of 1981, all new chemical substances that were produced or imported in the EU had to undergo a strict risk-assessment process. Under REACH, the burden of proof for demonstrating the safe use of chemicals was transferred from EU member states to industry. In effect, beginning in June 2008 and before the deadline of June 2018, all firms that manufactured or imported more than one metric ton of a chemical substance per year were required to register the chemical in a central database. This meant that all chemical substances produced before 1981 did not need to undergo the safety-assessment process again, and that firms that produced or imported less than one ton of a given chemical per year did not need to register these substances. The REACH provisions had been designed to be phased in gradually over a period of eleven years because of the complexity of the process and because of the need to find replacements for some chemicals that were considered too dangerous to be retained. As stated previously, REACH is a general regulatory framework, and it does not apply specifically to nanosubstances. Critics of the law said that as most nanochemicals were so small, they were produced in quantities that were below one ton per year, and this meant that they went unregulated.

They also criticized the fact that there were no specific provisions for nanosubstances. Under the Belgian presidency of the EU in 2010, there had been a call for the creation of a specific nanomaterials registry under REACH, but the initiative had not produced any concrete results as of mid-2011.

The 2011 PEN inventory of nanobased consumer products on the market provided an estimate of their major components (see PEN, Project on Emerging Nanotechnologies, http://www.nanotechproject.org/inventories/consumer/, accessed March 2011). It found that out of a total of 1,317 products or product lines containing nanoparticles, nanosilver was the major component of 313 of them, followed by nanocarbon and nanotitanium. Most of these products, whose appeal for customers relied on entirely new properties, were largely unregulated, and there was no comprehensive data concerning the actual quantities of the nanoparticles produced.

There were in mid-2010 only two cases of mandatory regulation applying specifically to nanotechnologies in the world. The first one was a municipal ordinance of the city council of Berkeley, California, in the U.S. It amended existing health and safety rules to demand a full toxological report from all facilities manufacturing nanoparticles:

> All facilities that manufacture or use manufactured nanoparticles shall submit a separate written disclosure of the current toxicology of the materials reported, to the extent known, and how the facility will safely handle, monitor, contain, dispose, track inventory, prevent releases and mitigate such materials. (Berkeley City Council, The Manufactured Nanoscale Health and Safety Ordinance [2006], § 15.12.040

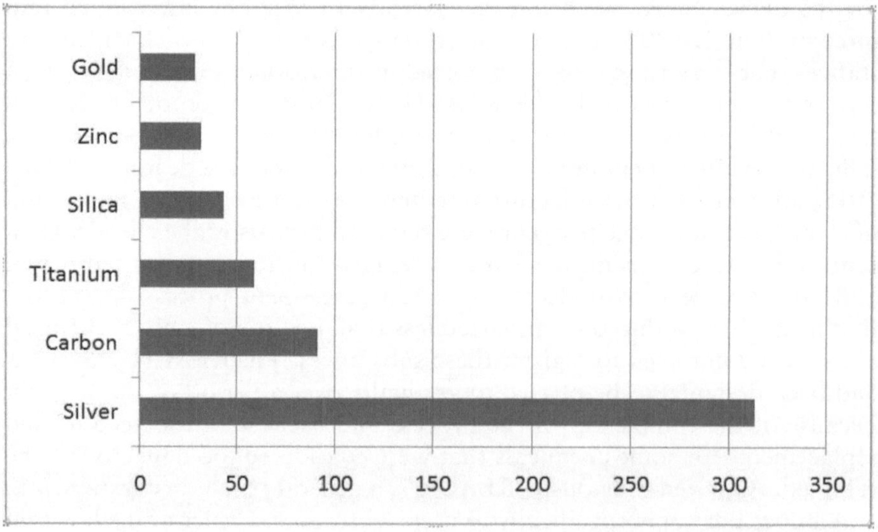

Figure 24.1 Major materials present in nanobased products on the market in 2011. *Source:* PEN, 2011. Author's chart.

There were rumors that this initiative would spread to other U.S. city councils, and the city of Cambridge in Massachusetts was reported to be next on the list. However, in July 2008, a report to the Cambridge city manager, *Recommendations for a Municipal Health & Safety Policy for Nanomaterials,* concluded that in recognition of the limited health-effects data and the absence of a clear consensus on best practices and standards for engineered nanomaterials, the Cambridge Public Health Department, in collaboration with the Cambridge Nanomaterials Advisory Committee, *did not recommend* that the city council enact a new ordinance regulating nanotechnology at that time (Cambridge Nanomaterials Advisory Committee and Cambridge Public Health Department 2008).

The second piece of mandatory regulation was a text from the European Parliament (2009)—Regulation 2009/1223/EC. It required the manufacturers of new cosmetic products containing nanomaterials to notify the European Commission and to provide certain information six months before the product was placed on the European market. The limitations of this piece of legislation were that it only concerned cosmetics, and that it only concerned new products. For example, it did not apply to all the nanotechnology-based cosmetics that had been put on the market prior to 2009 (approximately 140 products, according to the PEN inventory).

Apart from these two examples of mandatory regulation, all other existing forms of specific nanotechnology regulation were voluntary schemes. They could be divided into three broad categories: voluntary reporting schemes, codes of conduct, and standards.

VOLUNTARY REPORTING SCHEMES

The Voluntary Reporting Scheme (VRS) for Manufactured Nanomaterials was set up in the UK in September 2006 as a temporary experiment with an expected life of two years. It was run by Defra, the Department for Environment, Food and Rural Affairs. The scheme aimed to provide an indication of the kinds of nanomaterials currently in development and production to help inform policy-making decisions and to focus efforts and funding on areas that were relevant to the UK's current nano manufacturing and research base. The program asked for data that could be provided on manufactured nanomaterials from anyone involved in the manufacture or use of engineered nanomaterials, or from anyone involved in nanoscience research or in managing wastes consisting of engineered nanoscale materials. The program included regular updates to assess its implementation (Department for Environment, Food and Rural Affairs 2008).

In 2008, the seventh such quarterly report indicated that after twenty-two months, a total of eleven submissions had been received since the scheme's launch in September 2006—nine from industry, and two from academia. These appeared to be quite poor results in view of the UK's

estimated nanotechnology production. For example, the Qineti Group—said to be the UK's first production facility dedicated to the high-volume production of specialist nanomaterials—stated on its website that it had two production rigs that were each capable of producing up to several kilos of material an hour, and that it was looking at over twenty-five key nanomaterial projects (retrieved from:
http://www.qinetiq.com/home/newsroom/news_releases_homepage/2003/1st_quarter/nanomaterials0.html, accessed July 2007).

In January 2008, the United States Environmental Protection Agency (EPA) launched the Nanoscale Materials Stewardship Program (NMSP), a similar voluntary reporting scheme covering engineered nanoscale materials manufactured or imported for commercial purposes (Environmental Protection Agency 2007). The NMSP program invited interested parties to participate in a 'basic' program by submitting existing data on the engineered nanoscale materials they manufactured, imported, processed, or used. The EPA also invited interested parties to participate in an 'in-depth' program to test engineered nanoscale materials they manufactured, imported, processed, or used. This scheme seemed to meet the same fate as the UK Voluntary Reporting Scheme. When it launched the NMSP program, the EPA said it expected to receive 240 submissions from 180 companies under the basic program, and to attract 15 participants in the in-depth program. This projection was based on an estimate that in 2005 more than 600 companies were manufacturing and applying nanotechnology. However, in 2008, the EPA provided on its website a list of only nine companies that had made submissions and eleven companies that intended to, with two more volunteering for the 'in-depth' program component.

So it appeared that the voluntary reporting schemes did not meet with industry enthusiasm. Interviews with industry representatives suggested that the major reason for this failure was that when a company attempted to give an earnest report of its nanoactivities, the very information it provided became a source of scrutiny—and sometimes criticism or concern. Companies felt that because of the current lack of data, the data *they* provided became the source of a benchmarking of nanotechnology risk evaluation for which they were accountable because they were the source of the data.

CODES OF CONDUCT

At approximately the same time, a few initiatives were launched to design codes of conduct founded on voluntary implementation. One was the product of a public institution: the EU *Code of Conduct for Responsible Nanosciences and Nanotechnologies Research* (European Commission 2008). Based on the precautionary principle, this voluntary code covered seven general principles, including sustainability, precaution, inclusiveness, and accountability. Its main goal was to help research institutes, universities, and companies in the EU ensure the safe development and use of nanotechnologies in the face of knowledge gaps and uncertainties about the future impact of these technologies on human health and the environment. The EU launched a period of public consultation from October 2009 to January 2010 to evaluate the impact of this voluntary initiative and the revisions that might be introduced. The public consultation had been opened to all stakeholders directly or indirectly involved or interested in nanosciences and nanotechnologies research, but very little feedback was received. Only forty-nine valid answers were received from seventeen countries, including four that were outside the EU. As of mid-2011, there had not been an updated version of the code.

Whereas the EU *Code of Conduct for Responsible Nanosciences and Nanotechnologies Research* was an initiative coming from a public institution, the UK *Responsible NanoCode*, also released in 2008, was a joint production of several bodies including an asset-management firm (Insight Investment) and a trade association (Nanotechnology Industries Association), as well as a research institution (The Royal Society) (Royal Society, Insight Investment, and Nanotechnology Industries Association [NIA] 2008). The project had started in 2006 and aimed to explore the societal and economic impacts of the technical, social, and commercial uncertainties related to nanotechnologies. The idea had come from corporate codes of conduct that were beginning to be put in place by industry. The *Responsible NanoCode* had seven principles and aimed to establish a consensus of what constituted good practice across the nanotechnology value chain (i.e., from research and development to manufacturing, distribution, and retailing) so that businesses could align their processes with emerging good practices and form the foundation for the development of indicators of compliance. It illustrated expected behaviors and processes, not standards of performance, and it did not endeavor to be an auditable standard. The scope of its application was limited, and its 2008 update was not followed by any more recent documents (see http://www.responsiblenanocode.org/documents/TheResponsibleNanoCodeUpdateAnnoucement.pdf, accessed April 2011).

Another example of a voluntary tool of governance was the Nano Risk Framework, a coproduction of industry and of an NGO. In June 2005, the chemical company DuPont and the environmental group Environmental Defense Fund began developing this initiative. Their objective was to help answer questions an organization should consider in developing applications using nanomaterials, including providing a way to address areas of incomplete or uncertain information using reasonable assumptions and appropriate risk-management practices (Environmental Defense Fund and DuPont 2007). The framework included guidance on how to communicate information and decisions to stakeholders. One year after its launch, the Nano Risk Framework was available in French, Mandarin, and Spanish, and DuPont made compliance with the framework mandatory for all its nanotechnology work. However, after two interactive workshops on nanorisk management were held in 2008, the initiative seemed to peter out gradually.

Finally, there was a purely corporate initiative. The chemical company BASF, which had developed in 2004 the first corporate code of conduct outlining the company's duties to workers, investors, and clients, introduced in 2007 a specific code of conduct for nanotechnologies. Based on principles of responsible management, the code had four commitments detailing the company's approach, and included strong commitments to safety and transparency, which are evident in the following quotes: "Economic considerations do not take priority over safety and health issues and environmental protection," and "We disclose new findings to the authorities and the public immediately." Retrieved from http://corporate.basf.com/en/sustainability/dialog/politik/nanotechnologie/kodex/?id=mXz1wD0.gbcp*9S, accessed July 2007

The analysis of these voluntary codes reveals that they had a limited impact. They were designed as voluntary, principles-based guidelines establishing a consensus of what constituted good practices in businesses across the nanotechnology value chain in order that these businesses could align their processes with emerging good practices. However, they dealt more with management principles than with the technology risks themselves. They illustrated expected behaviors and processes—not standards of performance—and they did not aim to be auditable standards.

Also, these voluntary codes did not include tools for evaluating their successful (or not) implementation. It was therefore impossible to measure their effectiveness in terms of regulating the emergence of the nanoindustry.

STANDARDS

At the same time as all the initiatives just described were taking shape, there was a great deal of activity from standardization organizations working on nanotechnology standards. Djelic and Sahlin-Andersson (2006) and Brunsson and Jacobsson (2002) already pointed out that standardization was becoming a new transnational institution, and this was particularly true in the case of nanotechnologies.

It is commonly agreed that standards exist to

- support commercialization and market development;
- provide a basis for procurement; and
- support appropriate legislation/regulation.
  (*Source:* Phelps 2006)

There are two forms of initiatives, national initiatives and international ones. In terms of national initiatives with an interest in nanotechnologies, one can mention among others the American National Standards Institute

(ANSI), the Chinese National Nanotechnology Standardization Committee, the French Standards Agency (AFNOR), the UK British Standards Institute (BSI), and the Japanese Standards Association (JSA).

National standardization agencies share in common the view that standardization is a means to assert competitive advantage. For instance, the BSI stated on its website that "standardization has a key part to play in all innovative technologies, and nanotechnology is no exception." It considered that its nanoproject, the publication of nine documents for nanotechnology terminology and standardization guidance for UK industry, did not only support worker, public, and environmental safety, but that it also underpinned commercialization and procurement, and was part of BSI's overall initiative to support the innovation agenda of the UK government. (Retrieved from http://www.bsi-global.com/en/About-BSI/News-Room/BSI-News-Content/Disciplines/Nanotechnology/Nanotech-launch/, accessed March 2008.)

For the French AFNOR, standards were differentiation tools, and "standardization [was] an excellent method of international penetration for our competitiveness clusters" (Association Française de Normalisation [AFNOR] 2007).

The American National Standards Institute's Nanotechnology Standards Panel (ANSI-NSP) served as the cross sector coordinating body for the purposes of facilitating the development of standards in the area of nanotechnology including nomenclature/terminology, materials properties and testing, measurement, and characterization procedures. Its strategy was to establish a framework that could be used by all interested parties to "advance trade issues in the global marketplace, enhance consumer health and safety, meet stakeholder needs and, as appropriate, advance U.S. viewpoints in the regional and international arena." (Retrieved from *United States Standards Strategy*, http://www.ansi.org/standards_activities/nss/usss.aspx?menuid=3, accessed April 2008.)

Clearly, standards were perceived as playing a vital role in economic dynamism, and national standards agencies viewed standardization ownership—the fact that it was their standards which were applied by most in a given industry—as a way to assert a position of leadership.

What did this mean for nanotechnologies, when regulation was considered unsatisfactory or insufficient by many stakeholders (United Nations Secretariat 2009)?

On April 1, 2005, China became the first country in the world to establish national standards for nanoscale materials. These standards consisted of a list of seven items: four standards for nanoscale nickel powders, zinc oxide, titanium dioxide, and calcium carbonate; two standards for the testing of the specific surface area and pore-size distribution of powdered or solid materials via gas adsorption; and one standard for the granularity of nanosized powders. According to ACON AG, a leading science and business consultancy focusing on Asia, in doing so, China gained a competitive advantage, as these standards were expected to lay a good foundation for market access, market regulation, and proper application of nanomaterials globally.

Shortly thereafter, the International Organization for Standardization (ISO) launched a technical committee on nanotechnology, ISO/TC 229, in May 2005. Keeping up with these movements, the Japanese Ministry of Economy, Trade and Industry (METI) launched a panel within the Japan Standards Association called the Nanostandardization Panel. It developed into the Council on Nanotechnology Standards in Japan in September 2005 and hosted the second plenary meeting of the ISO/TC 229 in June 2006. In June 2005, China launched its own National Technical Committee on nanotechnology standardization, with the approval of SAC (Standardization Administration of the People's Republic of China): SAC/TC 279.

By 2007, each of the projects of ISO/TC 229 had been categorized into one of the technical committees' working groups:

WG 1 - Terminology and Nomenclature, convened by Canada
WG 2 - Measurement and Characterization, convened by Japan
WG 3 - Health, Safety, and Environment (convened by the USA)

ISO TC229 MEETINGS		
Meeting	*Date*	*Place*
1st plenary	Nov-05	United Kingdom
2nd plenary	Jun-06	Japan
3rd plenary	Dec-06	Republic of Korea
4th plenary	Jun-07	Germany
5th plenary	Dec-07	Singapore
6th plenary	May-08	France
7th plenary	Nov-08	PR China
8th plenary	Jun-09	USA

Figure 24.2 ISO/TC 229 meetings.
Source: author.

In October 2007, China's national standards agency SAC submitted two new work-item proposals addressing specifications for nanomaterials in terms of possible applications. In response to these newly identified needs, a new working group on Material Specifications (WG 4) was formed in early 2008, convened by China.

Figure 24.2 illustrates how global the process of nanotechnology standardization is. Over a period of five years, ISO meetings for TC 229 on nanotechnologies were convened in ten countries, and the ninth plenary session gathered over 140 representatives from thirty-two member countries. The table also shows the importance of Asia in this process—half of the meetings are held in Asia.

Specific tasks of ISO/TC 229 include developing standards for

- terminology and nomenclature
- metrology and instrumentation, including specifications for reference materials
- test methodologies
- modeling and simulation
- science-based health, safety, and environmental practices (International Organization for Standardization 2007)

As of September 2010, five international standards had been published by ISO/TC 229, and nearly forty were under development.

The five published standards were as follows:

1. ISO/TS 27687:2008. Nanotechnologies—Terminology and definitions for nanoobjects—Nanoparticle, nanofibre and nanoplate (published August11, 2008). ISO/TS 27687:2008 listed unambiguous terms and definitions related to particles in the field of nanotechnologies. It was intended to facilitate communication between organizations and individuals in industry and those who interacted with them.
2. ISO/TR 12885:2008. Nanotechnologies—Health and safety practices in occupational settings relevant to nanotechnologies (published September 30, 2008). ISO/TR 12885:2008 described health and safety practices in occupational settings relevant to nanotechnologies. It focused on the occupational manufacture and use of engineered nanomaterials.
3. ISO/TS 80004–3:2010. Terms and definitions related to carbon nanoobjects in the field of nanotechnologies (published April 19, 2010). ISO/TS 80004–3:2010 listed terms and definitions related to carbon nanoobjects in the field of nanotechnologies. It was intended to facilitate communication between organizations and individuals in industry and those who interact with them.
4. ISO/TR 11360:2010. Nanotechnologies—Methodology for the classification and categorization of nanomaterials (published July 12,

2010). ISO/TR 11360:2010 described a classifying system, termed a 'nanotree,' upon whose basis wide ranges of nanomaterials can be categorized, including nanoobjects, nanostructures, and nanocomposites of various dimensionality of different physical, chemical, magnetic, and biological properties.

5. ISO 29701:2010. Nanotechnologies—Endotoxin test on nanomaterial samples for in vitro systems—Limulus amebocyte lysate (LAL) test (published September 3, 2010). ISO 29701:2010 described the application of a test using Limulus amebocyte lysate (LAL) reagent for the evaluation of nanomaterials intended for cell-based in vitro biological test systems. The test was suitable for use with nanomaterial samples dispersed in aqueous media, e.g., water, serum, or reaction medium, and with such media incubated with nanomaterials for an appropriate duration at 37°C. ISO 29701:2010 was restricted to test samples for in vitro systems, but the methods could also be adapted to nanomaterials to be administered to animals by parenteral routes.

The elaboration of ISO standards is a fairly long process that includes six stages: proposal stage, preparatory stage, committee stage, enquiry stage, approval stage, and publication stage. All standards are reviewed within three years of publication and every five years after the first review by all the ISO member bodies. In mid-2010, the nano standards were still at a preliminary stage, and no specific nano ISO certification that could assess consistent rules or guidelines of technical specifications was in sight. For example, ISO/TS 27687:2008 (concerning terminology and definitions), which had been published in 2008, had to undergo a revision process in September 2010. However, the intense activity and heavy international involvement in nano ISO working groups revealed this subject's importance for policy makers as well as for business stakeholders.

25 Interpreting Change

According to representatives of industries involved in bringing nanotechnology products to the market, nanoentrepreneurship is under high tension: Customers *want* innovation and new products. Nanoentrepreneurs strive to satisfy them as part of their strategy to develop new business activities. Yet they also know that if a product turns out to be damaging to people's health or the environment, they will have to pay the hard price for it. This fairly recent phenomenon has intensified strongly as part of environmental awareness. Nanoentrepreneurs do not want repeats of what happened with, for example, asbestos: U.S. studies showed that as of 2002, over 700,000 individuals had filed asbestos lawsuits, and the eventual cost of asbestos litigation has been estimated at between $200 and $265 billion (White 2004).

As a result, nanoentrepreneurs are increasingly acting as institutional entrepreneurs, seeking to create new institutions or to modify existing ones to suit their purposes (Maguire, Hardy, and Lawrence 2004). For instance, whereas until the end of the twentieth century technological emergence as well as the management of innovation tended to be entirely driven by developed countries (Aubert 2004), developing countries are today participating in the innovation and governance processes of nanotechnologies at various levels. As was underlined previously, Iran, India, and South Africa, among others, began organizing regular international nanotechnology events around 2005, and the database of the consultancy Nanowerk illustrates how generalized this phenomenon has become (see http://www.nanowerk.com/phpscripts/n_events.php, accessed April 2010).

Zucker et al. (2006) pointed out in their research on the rate of regional growth of new knowledge in nanotechnology that the latter is directly influenced by cross organizational collaborations and their associated flow of tacit knowledge. Similarly, by hosting these conferences—by giving a voice to the research of their own representatives—emerging nations modify the process of institutional emergence, i.e., the birth of new institutions or the modification of existing ones, as well as the governance of innovation.

These new developments are part of what I call the emergence of a new form of technological democracy. It is a new form of institutional dynamics whereby decision making is no longer mostly top down and developed-country-driven as it used to be but is becoming a more complex, globalized, and multistakeholder process. In this new form of international governance, national decisions or road maps on regulatory frameworks are likely to be superseded by new forms of international regulation that are being put into place whether individual nation-states want them or not. Indeed, a traditional top-down approach of regulation is no longer considered sufficient in addressing the concerns raised by emerging technologies (Irwin, Jones, and Stilgoe 2006). At the same time, there is strong disagreement whether existing regulatory frameworks can actually cover their development (United Nations Secretariat 2009; Hodge, Bowman, and Ludlow 2009). Besides, there is no consensus on the value of the attempts at voluntary regulation. It seems that the voluntary reporting schemes that some governments were trying to put in place, such as the EPA's Nanoscale Materials Stewardship Program (NMSP) in the US or the UK's Voluntary Reporting Scheme for Manufactured Nanomaterials, did not meet industry's approval.

In this regulatory disorder of competing institutional logics, the midterm solution for all parties seemed to turn to voluntary regulation in the form of international standards and certification. It would be much cheaper to use government-based frames of voluntary reporting schemes. Indeed, the application of standards and certification is expensive because it requires the creation and conduction of costly tests to demonstrate compliance with agreed standards or best-practice scenarios. Yet it may become the favored solution for several reasons.

First, international standards are global. This is particularly significant because the development of nanotechnologies pertains to a transnational system of governance. It is research intensive (European Commission 2007; Roco 2006) and, apart from public institutions, the largest R&D facilitators are multinational corporations that have the most resources. As a consequence, the development of new nanobased activities—through research and seed funding as well as through R&D agreements—has close links with large multinational corporations (MNCs) and, therefore, with their transnational system of governance (Larédo et al. 2009).

In addition, more than being the source of entirely new sectors of business activity, nanotechnologies tend to diffuse in existing products and services. Indeed, nanoproducts are primarily used to enhance the qualities or specific characteristics of existing products and services. This mode of diffusion means that the development of new nanobased business activities mainly happens within mature organizations (OECD 2009; Larédo et al. 2009). This also suggests embeddedness in the systems of governance of large multinational companies.

Finally, studies of nanoentrepreneurship mainly analyze it in regional terms, which is a transnational approach. They may have different appreciations of the regional share of nanoentrepreneurship in the global market, especially between the three blocs of Asia, the Americas, and Europe

(OECD 2009; Chen and Roco 2008). Yet the early development of nanotechnologies coincided with globalization (Nanoforum 2005), and all agree that nanoentrepreneurship clearly targeted global markets from the start (Nanoforum, Malsch, and Oud 2004; PCAST 2010).

In the face of such a particular context, international standards have been devised in an international manner. They do not 'belong' to any particular entity, as national regulations do. The history of technology regulation is full of stories on the difficulties stemming from conflicting national regulatory frameworks for new technologies. The case of GMO disputes (Blaustein 2008) or that of stem-cell research (Mariani 2002) are only two examples of a well-known problem. Of course, critics argue that the mechanisms of international standardization are far from transparent and are influenced by the intense lobbying carried out by some nations to support their own interests, but on the whole, international standards represent the best available international regulatory framework: in fact, there is no other.

The second reason why international standards are particularly well adapted to frame the emergence of nanotechnologies is that they are generic, which corresponds well to the transtechnological quality of nanodevelopment.

Third, standards are certifiable, and this gives them a seal of expertise and objectivity that is reassuring in light of the high-potential, high-risk quality of nanodevelopment. In interviews with representatives of the nanotechnology and insurance industries, the desire of both sectors to avoid regulatory risks, i.e., risks linked with changes in regulation, was evident. It is clear that standards are helpful because they enable both sectors to act according to clear benchmarks. Their legitimacy stems from their global reach as much as from the fact that they introduce a workable framework in a regulatory void. All these facts come together to give them a competitive advantage, especially in the specific historical context that combines an increasing awareness of environmental issues, a lack of information concerning potential risks associated with nanotechnologies, and regulatory confusion in existing frameworks. It is unknown at this stage if they will ever supersede existing forms of regulation. Yet they do correspond to a loss of ownership of regulation, another sign of a rupture in institutional emergence.

26 Conclusion

The year 2009 marked both the bicentenary of Charles Darwin's birth and his book *On the Origin of Species* (Darwin [1859] 2003). One of Darwin's points was that living species evolved throughout the passing of time, and that this evolution reflected a propensity for survival that demanded adaptation to environmental stimuli and pressure. In his view, natural species evolve over time through a process of elimination that favors the best-adapted specimen, and this is precisely what shapes their evolution. Darwin built his theory from his observations, which took place over a period of several decades and derived from many geographical locations. He did not, however, approach the question of the evolution of mental constructs. It was too early, and there was still much work to be done with his general theory of the interrelatedness of different species, climate, and the environment. He did, though, write an opus on the expression of the emotions in man and animals (Darwin and Ekman 1998) to illustrate that there is no difference of *kind*, but only of degree, between man and other animals.

Darwin's point was that the gradation of change that separates man from other animals can be extended to all organic forms, and that this change is a consequence of the gradual environmental adaptation continuously and cumulatively operating over time. But there, he was focusing on the relationship between humans and animals in their expressions of emotions such as fear, anger, disdain, or pleasure. He did not really seek to elucidate if or how these expressions had evolved. More recently, scientists, and particularly psycho-scientists, have begun exploring how emotions and mental constructs evolve (Buss 1998; Tooby and Cosmides 2000). Popper and Kuhn sought for their part to investigate the advance of scientific knowledge. For Popper(2002), scientific theories, and human knowledge in general, are generated by the human mind in order to solve problems that have arisen in specific historic cultural settings. According to him, theories that better survive the process of 'refutation' are not more true but, rather, more 'fit'—in other words, more adapted to the problem considered. So the way scientific knowledge advances is through continuous interaction between tentative theories (conjectures) and error elimination (refutation). This process, which is very similar to the process of natural selection and genetic evolution, explains both the progress of human knowledge and why scientific knowledge advances toward solving more and more complex problems.

Kuhn(1962) had also written extensively on the history and philosophy of science. A contemporary of Popper, he had a completely different interpretation of the development of science. Whereas Popper's theory suggests that progress follows a linear evolution, Kuhn's own work shows that major developments in science are mostly revolutionary. One of his most important insights was that scientific progress is not a linear accumulation of new knowledge, but is instead the result of revolutions in which the nature of scientific inquiry within a particular field is abruptly transformed. For Kuhn, generally speaking, science emerges through three distinct stages. The first stage is 'prescience.' At this stage, the various components of knowledge to come are still scattered around without a central core, or 'paradigm.' Then comes the second stage, 'normal science.' At this stage, a consensus has emerged to accept the new paradigm, and subsequent research consists in applying known methods to solve the new types of problems raised. This is the stage that Kuhn calls 'puzzle solving.' Just as puzzles generally have predetermined solutions, in doing 'normal' research scientists aim to discover something that is known in advance. At this stage, if results do not conform to the central paradigm, it does not mean that the paradigm must be refuted, but rather that the researcher has made mistakes in his or her analysis. Yet, if too many anomalous results build up, then science reaches a crisis, and a new paradigm arises, which supersedes the old results and accounts for the anomalous results. This is what Kuhn terms 'revolutionary science.'

Kuhn's theory challenged the prevailing positivistic concept of science as a unitary and unidirectional process of accumulating facts about a preexisting world. Building on Kuhn's work, various scholars had their own approaches and interpretations of what came to be known as the theory of paradigm shifts. Callon and Latour explored the sociological dimension of the process of the institutionalization of science. When analyzing the chronological steps of the emergence of science and technology, they showed that the process of stabilization and ontology acquisition had little to do with the intrinsic qualities of a given theory, but rested rather on the strength and breadth of the scientists' social network. The scientists' social network was what facilitated the 'translation' stage' (Callon, Law, and Rip 1986; Latour 1988). As was previously mentioned in the introduction, for Callon et al., scientific knowledge emerges from disconnected facts through the settlement of a controversy concerning their validity. The settlement of the controversy itself results from the success of the translation phase, during which a common language of reference is created so that previously unrelated entities can be connected and seen in interaction. The process of translation confers unity and intelligibility to otherwise obscure and apparently disconnected entities. It allows the institutionalization process to move into the domain of public understanding. It rests on the strength of the scientists' network, which

Callon et al. define as an ensemble of meta-organizations, including human and nonhuman 'actants,' interacting together.

Drexler(1986) paid tribute to Popper's and Kuhn's evolutionary theories in his own account of the evolution of matter itself, in his groundbreaking analysis of the emergence of nanotechnologies. For Drexler, the evolution of species, the evolution of the human environment, and the evolution of our understanding of the world all reflect the evolution of matter, which follows a similar pattern as the one first described by Darwin ([1859] 2003). His book put forward several new concepts. One concept was that the common point in all sciences from different fields is matter itself, and that the way the emerging laws of physics account for the evolution of matter connects directly to the evolution of all sciences. This theory has received particular visibility with the focus placed on converging technologies, or NBIC (Nanotechnology, Biotechnology, Information technology, and Cognitive science). In 2001, the U.S. National Science Foundation and Department of Commerce sponsored a workshop that resulted in an extended report, in which convergent technologies were defined as

> the synergistic combination of four major 'NBIC' (nano-bio-info-cogno) provinces of science and technology, each of which is progressing at a rapid rate. (Roco and Bainbridge 2003, ix)

The report explained that the four technologies—nanotechnology, biotechnology, information technology, and cognitive science—had undergone significant changes and expansion over the past fifty years, and that their respective paths of development were more and more closely interwoven. Furthermore, thanks to the convergence of NBIC, one could for the first time in human history envisage

> a comprehensive understanding of the structure and behavior of matter from the nanoscale up to the most complex system yet discovered, the human brain. (Roco and Bainbridge 2003, 1).

So this new generation of science and technology experts, who some criticized as being more akin to science fiction writers than reliable scientists, considered that science and technology had reached a stage when society was beginning to understand the scope and possibilities of modifying matter at will by acting on its basic components—its atomic structure. For them, it was only a temporary lack of adequate tools that prevented an impact on human beings and their environment without limitations. What did this mean in terms of the evolution of mankind? If nanotechnologies could enable scientists to modify matter at will by acting on its basic components, then they had the potential to bring about radical change to our own existence and evolution. This in turn marks a complete change in our understanding of the world.

Darwin could make initial guesses on the evolution of the physical characteristics of species over thousands of years because he had fossils to work with. Yet there are no equivalents of fossils for mental constructs. For example, although *Homo erectus,* a human ancestor, emerged at least two million years ago, it is generally agreed that the oldest art objects found on Earth are some 75,000 years old. They were small-drilled snail shells found in Blombos Cave in South Africa. So the presence of human species predated by more than a million years the first testimonies of their mental constructs. Although interpretations of the real meaning of the South African discovery diverge, the beads provided evidence for early symbolically organized behavior (d'Errico et al. 2005).

Before the appearance of writing, it was very difficult to find clues for understanding the evolution of mental constructs. It is quite obvious that the mental constructs of a person living in 2011 are hugely different from those of a caveman, but how can this evolution be explained? Can we use the adaptation model proposed by Darwin? If we consider the evolution of mental constructs as part of the evolution of species, we can argue that it also responds to a survival instinct, and that the dominant thinking is a consequence of an adaptation to the gradual environmental change continuously and cumulatively operating over time. The process may be similar to the process of elimination that favors the best-adapted specimen. In this case, the shaping of mental constructs follows a logic of preservation of the human species. We can then go back to the model proposed by Callon(1986) and Latour (1988) to interpret the institutionalization of science and technology and adapt it to the development of nanotechnologies. As a reminder, their model rested on a chronological process that includes several steps:

- Random appearance of seemingly disconnected facts
- Translation
- Dispute over the validity of facts
- Stabilization and ontology acquisition through settlement of controversies

If the shaping of mental constructs follows a logic of preservation of the human species, and if mental constructs take shape through the Callon and Latour model, then we can say that the current rupture in the institutional emergence of nanotechnologies comes from the preservation instinct of humans. This preservation instinct leads them to shape the development of new technologies so that these will not destroy mankind. As discussed in the introduction, the appearance of nanotechnologies—that is, the scientific ability to modify the atomic structure of matter in order to obtain new properties—coincided historically with the scientific ability to measure the environmental parameters that led to a rising awareness of sustainable-development issues.

Historically, nanotechnology emergence also coincided with the universalization of the Internet and information technology. Therefore, contrary to previous technological innovations, the translation process was very broad and did not remain in the small sphere of a tech-savvy elite. It became almost immediately part of the public domain. This particular conjecture of the evolution of the human species was what created the turning point: it signified that the two steps of 'dispute over the validity of facts' and 'ontology acquisition through settlement of controversies' were also much more inclusive than was the case with previous technological innovations. They corresponded to what I called a 'reshaping of technological democracy.' This also meant that including the notion of sustainable development in the development of nanotechnologies became key to the survival of the technology itself.

Appendices

Nanotechnology and Sustainable Development Milestones

Date	Event
1959	In his speech "There Is Plenty of Room at the Bottom," U.S. scientist Richard Feynman described the potential of atomic engineering.
1972	First global environmental meeting: the Stockholm Conference on Human Development. Creation of UNEP, the United Nations Environment Programme, to address global environmental problems and to act as 'the world's environmental conscience.'
1981	U.S. scientist Eric Drexler published the first technical paper on molecular nanotechnology in the *Proceedings of the National Academy of Sciences.*
1987	Publication of *Our Common Future*, also called the Brundtland report. It had been commissioned in 1983 by the UN and other organizations and provided the first official definition of sustainable development: "Development that meets the needs of the present without compromising the ability of future generations to meet their own needs".
1989	First major conference on nanotechnology, run by the Foresight Institute in Palo Alto, California. A gathering of nanotechnology researchers predicted that scientists would one day achieve thorough control over the structure of matter down to the level of individual atoms and molecules. The conference proceedings became a book: *Nanotechnology, Research and Perspective.*
1992	The UN Conference on Environment and Development, also called the Rio Earth Summit, gathered more than 100 heads of state representing 179 national governments in Rio de Janeiro, Brazil. As of 1989 thousands of representatives of NGOs and civil society had started contributing to the discussion process leading to the 1992 conference. Signature of the Rio Declaration and of Agenda 21 in the wake of the UN-led Rio Earth Summit. Principles 15 and 16 of the Rio Declaration introduced the precautionary principle and the 'polluter pays' principle.
1996	South Africa drafted its new constitution. Section 24 stated that everyone had the right to an environment that was not harmful to their health or well-being, and to have the environment protected, for the benefit of present and future generations.

Continued

1997	Poland drafted its new constitution. Article 5 stated that the Republic of Poland (. . .) "shall ensure the protection of the natural environment pursuant to the principles of sustainable development."
1998	Signature of the Aarhus Convention on Access to Information, Public Participation in Decision-making and Access to Justice in Environmental Matters. In 2011, this legally binding document had been signed by over forty parties, including the European Commission. It forged a new process for public access to information on the environment held by public authorities, as well as for public participation in the negotiation and implementation of international agreements.
1999	Kraft Foods, a leading U.S. food and beverage firm, launched the industry's first nanotechnology food laboratory. In 2000, Kraft launched the NanoteK consortium, providing funding for fifteen universities and national research laboratories to conduct basic research for nanofood-technology.
2000	The EU wrote a complete text of reference concerning the precautionary principle called the *Communication from the Commission on the precautionary principle*, COM (2000) 1 "(Commission of the European Communities 2000). This then served as a basis for all EU regulation.
2001	The Japanese Mitsubishi Corporation launched the world's first nanotechnology private equity fund called Nanotech Partners. The equity fund focused exclusively on nanotechnology around the world.
	Launch of the National Nanotechnology Initiative (NNI) by the U.S. Congress, a program established to coordinate activities in nanoscale science, engineering, and technology at the federal scale.
2002	Launch of Nanoforum by the European Commission, a Pan-European information portal and thematic network on nanotechnology aimed at strengthening the EU's economic competitiveness in this field.
2003	The first World Nano Economic Congress (WNEC) took place in Washington, D.C., U.S., to assess the impact of nanotechnology on business.
	Sequoia Pacific Research Company, a chemical research and development firm founded in 2001 that specialized in nanobased chemical-process applications, produced a solution intended to prevent erosion. The solution was sprayed on 1,400 acres at Taos Pueblo American Indian land in the U.S. after a fire. It was the first large-scale release of a nano-based product in the environment.
	Launch of the Silver Nano (Silver Nano Health System) trademark by Samsung, Korea. Silver Nano was the name of an antibacterial technology that used silver nanoparticles in washing machines, refrigerators, air conditioners, air purifiers, and vacuum cleaners in order to kill germs and bacteria.
	European Parliament's Committee on Industry, External Trade, Research and Energy (ITRE) set up research on nanoregulation and nanosafety (part of the EU's 6th Framework Programme, FP6).
	U.S. Congress passed S. 189, the 21st Century Nanotechnology Research and Development Act (U.S. Congress, 2003).

2004	According to the first toxicity study of engineered nanoparticles in aquatic species, juvenile largemouth bass living in aquariums contaminated with water-soluble carbon nanoparticles developed brain damage and other signs of physiological distress. (Oberdorster, 2004). That same year, at the first scientific conference on nanotoxicity, Nanotox 2004, initial findings from research on rats illustrated that gold nanoparticles could move across the placenta from mother to fetus.
	France added an Environmental Charter to its constitution. Article 1 stated that everyone had the right to live in a balanced environment that showed due respect for health. Article 4 stated that "everyone shall be required, in the conditions provided for by law, to contribute to the making good of any damage he or she may have caused to the environment."
	The European Commission published *Towards a European Strategy for Nanotechnology*. COM (2004) 338. The objective of the text was to bring the discussion on nanoscience and nanotechnology to an institutional level and to propose an integrated and responsible strategy for Europe.
	First meeting of the American National Standards Institute's Nanotechnology Standards Panel (ANSI-NSP). ANSI-NSP served as the cross- sector coordinating body for the purposes of developing standards in the area of nanotechnology, including terminology, materials properties, testing, measurement, and characterization procedures.
	Applied Digital, on behalf of its subsidiary VeriChip (renamed Positive ID Corporation in 2009), received authorization from the U.S. Food and Drug Administration (FDA) to sell chips for human implantation. In 2005, RFID chips were implanted in humans, to store medical data, or for 'hobbyists' who used them to open doors or to switch on lights or computers.
2005	Launch of *Nanosciences and Nanotechnologies: An Action Plan for Europe 2005–2009*, COM (2005) by the European Commission. It was a European road map for "immediate implementation of a safe, integrated and responsible strategy for Nanosciences and Nanotechnologies (N&N)."
	Controversy over whether Samsung's Silver Nano washing machine could prove harmful to the environment if effluent from waste water could contain nanosilver and subsequently kill beneficial bacteria and disrupt ecosystem functioning.
	Launch of the Lux Nanotech Index in conjunction with the American Stock Exchange, by Lux Research Inc. The Lux Nanotech Index was not an investable product, but was meant to serve as a tool that investors could use to understand how the stock performance of companies involved in nanotech compared with the market as a whole.
	China was the first country in the world to establish national standards for nanoscale materials.
	ISO, the International Organization for Standardization, launched a technical committee on nanotechnology: ISO/TC-229.

Continued

2006	Global Crown Capital, an investment firm, launched what it believed to be the first nanotechnology hedge fund and venture fund in the U.S. The fund focused on companies in a broad range of industries involved in nanotech-related activities. The industries included health care, drug development, defense, chemicals, and electronics, among others.
	The U.S. Environmental Protection Agency (EPA) ruled that the Samsung silver ion–generating washing machine, which released nanosilver ions into wash water, was subject to registration requirements under the Federal Insecticide, Fungicide, and Rodenticide Act (FIFRA). The agency's decision that if a product "incorporates a substance intended to prevent, destroy or mitigate pests, it is considered a pesticide and is required to be registered" was a reversal of an earlier determination that said that the Samsung washer was a device, rather than a pesticide, and therefore not subject to regulation. Under the registration requirement, manufacturers must provide evidence that the use of nanosilver would not cause harm to public health.
2008	Soil Association Certification Ltd, the UK's largest organic certification body, said it had become the first organization in the world to ban man-made nanomaterials from its certified organic products. The products included cosmetics, food, and clothing, and Soil Association claimed that the new technology posed a serious threat to human health.
	The British Standards Institute (BSI) published the first publicly available extensive guide to safe handling of nanomaterials: *PD 6699-2:2007 Nanotechnologies - Part 2: Guide to Safe Handling and Disposal of Manufactured Nanomaterials.*
	First session of the OECD Intergovernmental Forum on Chemical Safety especially on nanotechnologies (Dakar, Senegal): "Nanotechnologies at the OECD."
2009	First public hearing on nanotechnology held by the U.S. Consumer Product Safety Commission (CPSC). The CPSC, with a staff of fewer than 400 employees, oversaw the safety of 15,000 types of consumer products.
	Introduction of the first international law specifically designed for nanotechnologies: Regulation 2009/1223/EC No 1223/2009 of the European Parliament and of the Council of November 30, 2009, on cosmetic products.
2010	Research on plastic nanoparticles confirmed that nanoparticles could cross the human placenta (Wick et al. 2010).
	As part of the process of a proposal of regulation for novel foods, members of the European Parliament restated their position to exclude foods produced by nanotechnology processes from the EU authorization list until they had undergone a specific risk assessment regarding their possible impact on health.
	Proposal by U.S. Senator Pryor on the Nanotechnology Safety Act of 2010 to amend the Federal Food, Drug, and Cosmetic Act to establish a nanotechnology program

	Grenelle 2: Law on the National Commitment to the Environment was passed in France. As of 2011, firms had to disclose an assessment by an independent third party on the social and environmental impacts of their activities. Firms could be held accountable for environmental damage produced by their activities, even from subsidiaries.
2011	The city of Bangalore opened India's first nanopark, intended to host industries related to nano science and technology.
	Eleven *international* nanotechnology standards had been published by ISO/TC 229, and nearly forty were under development. Published standards concerned terminology; definitions; health and safety practices in occupational settings; methodology for the classification and categorization of nanomaterials; and tests.

Chronology of Nanotechnology Involvement by the ETC Group: 2001–1007

DATE	EVENT	INSTITUTION	MEANS of OUTREACH	COMMENTS
January 2001	Publication of "The ETC Century" by Pat Mooney	Dag Hammarskjöld Foundation	Hard copy Internet	
June 2001	International seminar: "What Next? Technological Transformation: Addressing Nanotechnology And and Other Emerging Technologies in the ETC Century"	Dag Hammarskjöld Foundation	Venue in Europe Audience and Internet coverage	
September 2001	Regional seminar, follow-up of "What Next" international seminar	ETC	Venue in Asia (Thailand) Audience and Internet coverage	Event hosted by civil-society partners in the region
December 2001	Region seminar, follow-up of "What Next" international seminar	ETC	Venue in Latin America (Chile) Audience and Internet coverage	Event hosted by civil-society partners in the region

Continued

June 2002	Report on health and safety impacts of nanotechnology asks for a moratorium on engineered nanoparticles: *No Small Matter!*	ETC	Hard copy Internet	
December 2002	Regional seminar, follow-up of "What Next" international seminar	ETC	Venue in South Africa Audience and Internet coverage	Event hosted by civil-society partners in the region
January 2003	Report: *The Big Down: From Genomes to Atoms: Technologies Converging at the Nano-scale* (80-page report)	ETC	Hard copy Internet	First comprehensive analysis of the implications of nanoscale technologies by civil society
April 2003	Study explaining the case for a global moratorium on nanotech and including a scientific literature review: *Size Matters*	ETC Coauthored with Dr. Vyvyan Howard of the Developmental Toxico- Pathology Unit of the University of Liverpool. In 1999, Dr. Howard, as president of the Royal Microscopy Society, co-edited the first collection of papers (1999) to examine the toxicity of nanoparticulates.	Hard copy Internet Scientific community	
June 2003	ETC Group speaker at European seminar on nanotechnology and converging technologies at the European Parliament, Brussels			

September 2003	ETC identifies first major environmental release of nanotechnology product in an Indian territory in New Mexico (SoilSET TM soil binders, a revolutionary new nanotechnology product used for dust control, soil stabilization, revegetation, and erosion control caused by wind, rain, or water runoff).	ETC	Media Coverage Internet	Environmental release in New Mexico. Contacts with Indian war chief.
March 2004	Dinner with Prince Charles and members of the Royal Society to offer advice on nanotechnology	ETC	Media Coverage Internet	
November 2004	Report: *Down on the Farm*, on the impact of nanoscale technologies on food and agriculture	ETC	Media Coverage Internet	
January 2005	ETC workshops on new technologies at Porto Alegre (Brazil) World Social Forum	ETC	Media Coverage Internet	
June 2005	ETC Group hosted seminar series on nanoscale technologies in Geneva geared towards civil-society representatives	ETC, World Council of Churches, UN, South Center, and representatives of thirty international NGOs	Civil-society multiplier effect Media coverage-Internet	
2005	Publication of three reports on nanotechnology and intellectual property at key dates, including the opening of the GlenEagles G8 summit	ETC	Media Coverage Internet	

Continued

October 2006	ETC Group testifies at U.S. Food & Drug Administration's public meeting on nanotechnology	ETC, U.S. FDA	Media Coverage Internet	
January 2007	ETC launches Nano Hazard symbol contest. Winning symbol to be sent to major regulating agencies.	ETC	Media coverage (Winners of Nano Hazard symbol contest announced at World Social Forum, Nairobi, Kenya) Internet	Interactive action Reaches all sorts of stakeholders
April 2007	Open letter to the international nanotechnology community to reject proposed framework put forward by DuPont and Environmental Defense Fund	ETC and coalition of civil-society and labor organizations	Media Coverage Internet	

Communication from the Commission on the Precautionary Principle

COMMISSION OF THE EUROPEAN COMMUNITIES
Brussels, 02.02.2000
COM(2000) 1

SUMMARY

1. The issue of when and how to use the precautionary principle, both within the European Union and internationally, is giving rise to much debate, and to mixed, and sometimes contradictory views. Thus, decision-makers are constantly faced with the dilemma of balancing the freedom and rights of individuals, industry and organisations with the need to reduce the risk of adverse effects to the environment, human, animal or plant health. Therefore, finding the correct balance so that the proportionate, non-discriminatory, transparent and coherent actions can be taken, requires a structured decision-making process with detailed scientific and other objective information.
2. The Communication's fourfold aim is to:

   - outline the Commission's approach to using the precautionary principle,
   - establish Commission guidelines for applying it,
   - build a common understanding of how to assess, appraise, man age and communicate risks that science is not yet able to evaluate fully, and
   - avoid unwarranted recourse to the precautionary principle, as a disguised form of protectionism.

   It also seeks to provide an input to the ongoing debate on this issue, both within the Community and internationally.

3. The precautionary principle is not defined in the Treaty, which prescribes it only once—to protect the environment. But *in practice*, its scope is much wider, and specifically where preliminary objective scientific evaluation, indicates that there are reasonable grounds for concern that the potentially dangerous effects on the *environment, human, animal or plant health* may be inconsistent with the high level of protection chosen for the Community.

   The Commission considers that the Community, like other WTO members, has the right to establish the level of protection—particularly of the environment, human, animal and plant health,—that it deems appropriate. Applying the precautionary principle is a key tenet of its policy, and the choices it makes to this end will continue to affect the views it defends internationally, on how this principle should be applied.

4. The precautionary principle should be considered within a structured approach to the analysis of risk which comprises three elements: risk assessment, risk management, risk communication. The precautionary principle is particularly relevant to the management of risk.

   The precautionary principle, which is essentially used by decision-makers in the management of risk, should not be confused with the element of caution that scientists apply in their assessment of scientific data.

   Recourse to the precautionary principle presupposes that potentially dangerous effects deriving from a phenomenon, product or process have been identified, and that scientific evaluation does not allow the risk to be determined with sufficient certainty.

   The implementation of an approach based on the precautionary principle should start with a scientific evaluation, as complete as possible, and where possible, identifying at each stage the degree of scientific uncertainty.

5. Decision-makers need to be aware of the degree of uncertainty attached to the results of the evaluation of the available scientific information. Judging what is an "acceptable" level of risk for society is an eminently *political* responsibility. Decision-makers faced with an unacceptable risk, scientific uncertainty and public concerns have a duty to find answers. Therefore, all these factors have to be taken into consideration.

In some cases, the right answer may be not to act or at least not to introduce a binding legal measure. A wide range of initiatives is available in the case of action, going from a legally binding measure to a research project or a recommendation.

The decision-making procedure should be transparent and should involve as early as possible and to the extent reasonably possible all interested parties.

6. Where action is deemed necessary, measures based on the precautionary principle should be, *inter alia*:

   - *proportional* to the chosen level of protection,
   - *non-discriminatory* in their application,
   - *consistent* with similar measures already taken,
   - *based on an examination of the potential benefits and costs* of action or lack of action (including, where appropriate and feasible, an economic cost/benefit analysis),
   - *subject to review,* in the light of new scientific data, and
   - *capable of assigning responsibility for producing the scientific evidence* necessary for a more comprehensive risk assessment.

   Proportionality means tailoring measures to the chosen level of protection. Risk can rarely be reduced to zero, but incomplete risk assessments may greatly reduce the range of options open to risk managers. A total ban may not be a proportional response to a potential risk in all cases. However, in certain cases, it is the sole possible response to a given risk.

   Non-discrimination means that comparable situations should not be treated differently, and that different situations should not be treated in the same way, unless there are objective grounds for doing so.

   Consistency means that measures should be of comparable scope and nature to those already taken in equivalent areas in which all scientific data are available.

   Examining costs and benefits entails comparing the overall cost to the Community of action and lack of action, in both the short and long term. This is not simply an economic cost-benefit analysis: its scope is much broader, and includes non-economic considerations, such as the efficacy of possible options and their acceptability to the public. In the conduct of such an examination, account should be taken of the general principle and the case law of the Court that the protection of health takes precedence over economic considerations.

   Subject to review in the light of new scientific data, means measures based on the precautionary principle should be maintained so long as scientific information is incomplete or inconclusive, and the risk is still considered too high to be imposed on society, in view of chosen level of protection. Measures should be periodically reviewed in the light of scientific progress, and amended as necessary.

   Assigning responsibility for producing scientific evidence is already a common consequence of these measures. Countries that impose a prior approval (marketing authorization) requirement on products that they deem dangerous *a priori* reverse the burden of proving injury, by treating them as dangerous unless and until businesses do the scientific work necessary to demonstrate that they are safe.

Where there is no prior authorisation procedure, it may be up to the user or to public authorities to demonstrate the nature of a danger and the level of risk of a product or process. In such cases, a specific precautionary measure might be taken to place the burden of proof upon the producer, manufacturer or importer, but this cannot be made a general rule.

Notes

NOTES TO THE INTRODUCTION

1. A42/427, chap. 2, p. 54.
2. Chap. 1, p. 10.
3. A 42/427, 1.2.13.

NOTES TO CHAPTER 1

1. The microscope's inventors were two Swiss researchers, Gerd Binning and Heinrich Roher.
2. The discovery of fullerenes, molecules of carbon atoms assembled in the shape of a cage, led to the Nobel Prize in Chemistry for Harold Kroto, Robert Curl, and Richard Smalley in 1996, and to the elaboration of carbon nanotubes.
3. Source: Nanoposts 2010, *The Global Market for Carbon Nanotubes to 2015*, 2nd ed., p. 140.

NOTES TO CHAPTER 3

1. A university spinout, also called spinoff, is a company that is being set up to turn a technological invention coming from university research into a business venture. Usually, universities retain some of the Intellectual Property (IP) rights when a spinout is created.

NOTES TO CHAPTER 4

1. Source: http://www.nanoisrael.org/player_govt.asp, accessed January 10, 2011.
2. Sources: Irish Council for Science, Technology and Innovation (ICSTI), and FORFAS.

NOTE TO CHAPTER 10

1. The oft-quoted figure of $3 billion over a thirteen-year period covers a wide range of scientific activities related to genomics, of which human genome sequencing represents only a small fraction. For more detailed information

concerning the Human Genome Project, see http://www.ornl.gov/sci/techresources/Human_Genome/project/budget.shtml.

NOTES TO CHAPTER 12

1. "Contained use": Directive 90/219/EEC; later replaced by Directive 98/81/EC. "Deliberate release": Directive 90/220/EEC; later replaced by Directive 2001/18/EC.
2. Federal Register. Volume 64. Number 98. Pages 22671–22672. May 21, 1999.
3. Ibid.

NOTES TO CHAPTER 13

1. "Traceability and Labeling of GMOs": Directive 1830/2003 EC. "Genetically Modified (GM) food and feed": Directive 1929/2003 EC.

References

AFSSET. 2008. Les Nanomatériaux: Sécurité au travail. Maisons-Alfort : AFSSET.

AFSSET. 2010. Évaluation des risques liés aux nanomatériaux pour la population générale et pour l'environnement. Maisons-Alfort : AFSSET.

Aghion, Philippe, and Peter Howitt. 1997. *Endogenous Growth Theory*. ambridge, MA: MIT Press.

Akrich, Madeleine, Michel Callon, and Bruno Latour, eds. 2006. *Sociologie de la Taduction: Textes Fondateurs*. Paris : Presses de l'Ecole des Mines.

Aldrich, H., and C. Zimmer. 1986. "Entrepreneurship through Social Networks." In *The Art and Science of Entrepreneurship*. 1st ed. Edited by D. Sexton and R. W. Smilor. Pensacola, FL: Ballinger Publishing.

Alexander, Carol. 2008. *Market Risk Analysis*. Vols.1–4. Hoboken, NJ: John Wiley & Sons.

Anand, N., and Brittany C. Jones. 2008. "Tournament Rituals, Category Dynamics, and Field Configuration: The Case of the Booker Prize." *Journal of Management Studies* 45 (6): 1036–60.

Anand, N., and Mary R. Watson. 2004. "Tournament Rituals in the Evolution of Fields: The Case of the Grammy Awards." *Academy of Management Journal* 47 (1): 59–80.

Anderson, Alison G. , Alan Petersen, Clare Wilkinson, and Stuart Allan. 2009. *Nanotechnology, Risk and CommunicationI*. Basingstoke: Palgrave Macmillan.

Association Française de Normalisation (AFNOR). 2007. *Activity Report 2007*. http://portailgroupe.afnor.fr/v3/pdf/activity-report.pdf.

Aubert, Jean-Eric. 2004. *Promoting Innovation in Developing Countries: A Conceptual Framework*. World Bank Institute.

Auplat, Claire. 2006. "Do NGOs Influence Entrepreneurship? Insights from the Developments of Biotechnologies and Nanotechnologies." *Society and Business Review* 1 (3): 266–79.

Baker, T., and Reed E. Nelson. 2005. "Creating Something from Nothing: Resource Construction through Entrepreneurial Bricolage." *Administrative Science Quarterly* 50 (3): 329–66.

Bansal, P., and K. Roth. 2000. "Why Companies Go Green: A Model of Ecological Responsiveness." *The Academy of Management Journal* 43 (4): 717–36.

Barley, Stephen R., and Pamela S. Tolbert. 1997. "Institutionalization and Structuration: Studying the Links between Action and Institution." *Organization Studies* 18 (1): 93–117.

Beck, Ulrich. 1992. *Risk Society: Towards a New Modernity*: Thousand Oaks, CA: Sage Publications.

Bennett, Peter, and Kenneth Calman, eds. 1999. *Risk Communication and Public Health*: New York: Oxford University Press.

Berg, Paul, David Baltimore, Sydney Brenner, Richard O. Roblin, and Maxime Singer. 1975. "Summary Statement of the Asilomar Conference on Recombinant DNA Molecules." Paper read at International Conference on Recombinant DNA Molecules, Asilomar, California, June.

Berkeley City Council. 2006. The Manufactured Nanoscale Heath and Safety Ordinance. Berkeley Municipal Code Section 15.12.040 (Hazardous Materials and Wastes Management).

Blanchflower, David G., and Andrew J. Oswald . 1998. "What Makes an Entrepreneur?" *Journal of Labor Economics* 16 (1): 26–60.

Blaustein, S. A. 2008. "Splitting Genes: The Future of Genetically Modified Organisms in the Wake of the WTO/Cartagena Standoff." *Pennsylvania State Environmental Law Review* 16 (2): 367–401.

Bonaccorsi, Andrea, and Thoma Grid. 2005. "Scientific and Technological Regimes in Nanotechnology: Combinatorial Inventors and Performance." Paper read at LEM Paper Series 2005/13.

Bosma, Niels, Jonathan Levie, Rachida Justo, Jan Lepoutre, and Siri Terjesen. 2010. *Global Entrepreneurship Monitor: 2009 Executive Report. GEM* available online at http://www.gemconsortium.org/about.aspx?page=pub_gem_global_reports. Accessed April 2010.

Bosso, Christopher J., ed. 2010. *Governing Uncertainty: Environmental Regulation in the Age of Nanotechnology.* Oxford: RFF Press.

Brundtland, Harlem G., and World Commission on Environment and Development, eds. 1987. *Our Common Future.* Oxford: Oxford University Press.

Brunsson, Nils , and Bengt Jacobsson. 2002. *A World of Standards.* Oxford: Oxford University Press.

Buss, David M. 1998. *Evolutionary Psychology: The New Science of the Mind.* Boston: Allyn & Bacon.

Callon, Michel. 1986. "Some Elements of a Sociology of Translation: Domestication of the Scallops and the Fishermen of St Brieuc Bay." In *Power, Action and Belief: A New Sociology of Knowledge?* Edited by J. Law. London: Routledge.

Callon, Michel, Pierre Lascoumes, and Yannick Barthe. 2001. *Agir dans un monde incertain—Essai sur la démocratie technique.* Paris: Edition du Seuil.

Callon, Michel, J. Law, and A. Rip, eds. 1986. "How to Study the Force of Science." In *Mapping the Dynamics of Science and Technology: Sociologies of Science in the Real World.* London: Macmillan.

Cambridge Nanomaterials Advisory Committee, and Cambridge Public Health Department. 2008. *Recommendations for a Municipal Health and Safety Policy.* Cambridge: Cambridge Public Health Department.

Campbell, Kurt M., Robert J. Einhorn, and Mitchell B. Reiss. 2004. *The Nuclear Tipping Point: Why States Reconsider Their Nuclear Choices.* Washington, D.C.: Brookings Institution Press.

Cantillon, Richard. 1952. « Essai sur la nature du commerce en général. » Paris: INED. First published as a text manusript in London by Fletcher Gyles under the title "Essay on the Nature of Trade in General." 1755.

Chataway, Joanna, Joyce Tait, and David Wield. 2006. "The Governance of Agro- and Pharmaceutical Biotechnology Innovation: Public Policy and Industrial Strategy." *Technology Analysis & Strategic Management* 18 (2): 169.

Chavas, Jean-Paul. 2004. *Risk Analysis in Theory and Practice.* San Diego, CA: Elsevier Academic Press.

Chen, Hsinchun , and Mihail C. Roco. 2008. *Mapping Nanotechnology Innovations and Knowledge: Global and Longitudinal Patent and Literature Analysis.* Springer.

Cientifica Ltd., http://www.cientifica.eu

Cité des sciences et de l'industrie. 2007. *Cahiers d'acteurs—Nanotechnologies: le point sur les débats, des orientations pour demain.* Paris: Ministère de l'enseignement supérieur et de la recherche, Ministère de l'industrie.

Clarysse, Bart, Mike Wright, Andy Lockett, Els Van de Velde, and Ajay Vohora. 2005. "Spinning Out New Ventures: A Typology of Incubation Strategies from European Research Institutions." *Journal of Business Venturing* 20 (2): 183–216

Collins, Orvis F., and David G. Moore. 1970. *The Organization Makers.* New York: Appleton.

Commission of the European Communities. 2000. *Communication from the Commission on the Precautionary Principle.* COM(2000) 1. Brussels.

Commission of the European Communities. 2000. *Towards a European Research Area.* COM(2000) 6. European Commission.

Cothern, C. Richard, ed. 1995. *Handbook for Environmental Risk Decision Making.* Boca Raton, FL: CRC Press.

Currall, Steven C., E. B. King, Neal Lane, J. Madera, and S. Turner. 2006. "What Drives Public Acceptance of Nanotechnology?" *Nature Nanotechnology* 1: 153–55.

d'Errico, Francesco , Christopher Henshilwood, Marian Vanhaeren, and Karen van Niekerk. 2005. "Nassarius Kraussianus Shell Beads from Blombos Cave: Evidence for Symbolic Behaviour in the Middle Stone Age." *Journal of Human Evolution* 48 (1): 3–24.

Dacin, Tina. 1997. Review of "Institutions and Organizations," by W. Richard Scott. *Administrative Science Quarterly* 42 (4): 821–24.

Darby, Michael R., Qiao Liu,, and Lynne G. Zucker. 2004. "High Stakes in High Technology: High-Tech Market Values as Options." *Economic Inquiry* 42 (3): 351–69.

Darby, Michael R., Lynne G. Zucker, and Andrew G. Wang. 2004. "Joint Ventures, Universities, and Success in the Advanced Technology Program." *Contemporary Economic Policy* 22 (2): 145.

Darwin, Charles. (1859) 2003. *On the Origin of Species*: London: John Murray. Reprint, New York: Signet Classics. Citations refer to the Signet edition.

Darwin, Charles, and Paul Ekman. 1998 *The Expression of the Emotions in Man and Animals, Definitive Edition.* 3rd ed. New York: Oxford University Press.

David, Kenneth, and Paul B. Thompson, eds. 2008. *What Can Nanotechnology Learn From Biotechnology? Social and Ethical Lessons for Nanoscience from the Debate over Agrifood Biotechnology and GMOs.* Academic Press.

Davis, Jason P., Kathleen M. Eisenhardt, and Christopher B. Bingham. 2009. "Optimal Structure, Market Dynamism, and the Strategy of Simple Rules." *Administrative Science Quarterly* 54: 413–52.

de-Holan, Pablo, Nelson Phillips, and Thomas B. Lawrence. 2004. "Managing Organizational Forgetting." *MIT Sloan Management Review* 45 (2): 45–51.

Denzin, Norman, and Yvonna Lincoln, eds. 2000. *Handbook of Qualitative Research.* 2nd ed. London: Sage.

Department for Environment, Food and Rural Affairs. 2008. *UK Voluntary Reporting Scheme for Engineered Nanoscale Nanomaterials.* http://www.defra.gov.uk/environment/nanotech/policy/pdf/vrs-nanoscale.pdf.

DiMaggio, Paul J. 1983. "State Expansion and Organizational Fields." In *Organization Theory and Public Policy.* Edited by R. H. Hall and R. E. Quinn. Beverly Hills, CA: Sage.

DiMaggio, Paul J. 1988. "Interest and Agency in Institutional Theory." In *Institutional Patterns and Organizations: Culture and Environment.* Edited by L. G. Zucker. Cambridge, MA: Ballinger.

DiMaggio, Paul J., and Walter W. Powell. 1983. "The Iron Cage Revisited: Institutional Isomorphism and Collective Rationality in Organizational Fields." *American Sociological Review* 48 (2): 147–60.

DiMaggio, Paul J., and Walter W. Powell, eds. 1991. *The New Institutionalism in Organizational Analysis*. Chicago: University of Chicago Press.
Djelic, Marie-Laure, and Kerstin Sahlin-Andersson, eds. 2006. *Transnational Governance: Institutional Dynamics of Regulation*. Cambridge: Cambridge University Press.
Dorado, Silvia. 2006. "Social Entrepreneurial Ventures: Different Values So Different Process of Creation, No?" *Journal of Developmental Entrepreneurship*, 11(4) 368–391.
Dowling, Ann. 2004. *Nanoscience and Nanotechnologies: Opportunities and Uncertainties*. Royal Society and Royal Academy of Engineering.
Drexler, Eric. 1986. *Engines of Creation: The Coming Era of Nanotechnology*. New York: Anchor Books.
Drexler, Eric. 1991. *Unbounding the Future: The Nanotechnology Revolution*. New York: Quill/William Morrow.
Drexler, Eric, and Richard Smalley. 2003. "Nanotechnology." *Chemical & Engineering News* 81 (48): 37–42.
Dubini, Paola, and Howard E. Aldrich. 1991. "Personal and Extended Networks Are Central to the Entrepreneurial Process." *Journal of Business Venturing* 6 (5): 305–13.
Dupuy, Jean-Pierre, and Françoise Roure. 2004. *Les nanotechnologies: éthique et prospective industrielle*. Conseil Général des Mines, Conseil Général des Technologies de l'Information.
Einsiedel, E. F., and L. Goldenberg. 2004. "Dwarfing the Social? Nanotechnology Lessons from the Biotechnology Front." *Bulletin of Science, Technology & Society* 24 (1): 28–33.
Environmental Defense, and DuPont. 2007. *Nanorisk Framework*. http://www.edf.org/documents/6496_Nano%20Risk%20Framework.pdf:104.
Environmental Protection Agency. 2007. Nanoscale Materials Stewardship Program and Inventory Status of Nanoscale Substances under the Toxic Substances Control Act. http://www.epa.gov/fedrgstr/EPA-TOX/2007/July/Day-12/t13558.htm.
Ernst & Young. 2003. *Beyond Borders: Ernst & Young's Global Biotechnology Report*. Ernst & Young.
ETC Group. 2002. "No Small Matter! Nanotech Particles Penetrate Living Cells and Accumulate in Animal Organs." *ETC Group Communiqué* 76: 8.
ETC Group. 2003. "Size Matters!" In *Occasional Papers Series* 7 (1).
ETC Group. 2005. ETC Group External Review Report. Presented to ETC Group and the Swedish International Biodiversity Program. 66.
European Commission. 2003. *Key Figures 2003–2004*.
European Commission. 2004. *Nanotechnologies: A Preliminary Risk Analysis on the Basis of a Workshop Organized in Brussels on 1–2 March 2004 by the Health and Consumer Protection Directorate General of the European Commission*.
European Commission. 2004. *Report from the Commission on Monitoring the Application of Community Law (2003) Overall Position*. Twenty-first annual report.
European Commission. 2004. *Towards a European Strategy for Nanotechnology*. COM(2004) 338.
European Commission. 2005. *Nanosciences and Nanotechnologies: An Action Plan for Europe 2005–2009*. COM(2005) 243.
European Commission. 2007. *Nanosciences and Nanotechnologies: An Action Plan for Europe 2005–2009*. COM(2007) 505 final. First implementation report, 2005–2007.
European Commission. 2007. *Progress Report on the Sustainable Development Strategy 2007*. European Commission.

European Commission. 2008. *Code of Conduct for Responsible Nanosciences and Nanotechnologies Research*. C(2008) 424 final.

European Commission. 2008. *EU Nanotechnology R&D in the Field of Health and Environmental Impact of Nanoparticles*. Compiled by Pilar Aguar and Jose Juan Murcia Nicolas, Unit G4, Nano and Converging Sciences and Technologies, Research DG, version: 28.

European Commission. 2009. *MEPs Call for Review on Nano Safety* ENDS (Environmental Data Services). 412: 35–35.

European Commission, and Nanoforum. 2004. *Outcome of the Open Consultation on the European Strategy for Nanotechnology*.

European Communities. 2010. *Directory of Community Legislation*. European Publications Office.

European Community. 1992. The Maastricht Treaty.

European Parliament. 2009. Regulation 2009/1223/EC of the European Parliament and of the Council of 30 November 2009 on cosmetic products (recast).

European Science and Technology Observatory. 2003. *Review of GMOs under Research and Development and in the Pipeline in Europe*. European Commision Joint Research Centre.

European Union. 2003. *Tomorrow's Technologies, Today*. Cited August 2003. http://www.cordis.lu/euroabstracts/en/august03/feature03.htm.

European Union. 2007. Regulation 2006/1907/EC of the European Parliament and of the Council of 18 December 2006 concerning the Registration, Evaluation, Authorisation and Restriction of Chemicals (REACH).

Feenberg, Andrew, and Alastair Hannay, eds. 1995. *Technology and the Politics of Knowledge*. Bloomington, IN: Indiana University Press.

Feynman, Richard. 1960. "There's Plenty of Room at the Bottom." *Engineering and Science* 23(5): 22–36.

Fiorino, Daniel J. 1990. "Citizen Participation and Environmental Risk: A Survey of Institutional Mechanisms." *Science, Technology & Human Values* 15 (2): 226–43.

Fischer, Frank. 1993. "Citizen Participation and the Democratization of Policy Expertise: From Theoretical Inquiry to Practical Cases." *Policy Sciences* 26: 165–87.

Fligstein, Neil. 1997. "Social Skill and Institutional Theory." *The American Behavioral Scientist* 40 (4): 397–405.

Foddy, William. 1993. *Constructing Questions for Interviews and Questionnaires*. HongKong: Cambridge University Press.

Forrest, David. 1989. "Regulating Nanotechnology Development." Foresight Institute.

Forrester, Jay. 1961. *Industrial Dynamics*. Waltham, MA: Pegasus Communications.

Freeman, Chris, and Carlotta Perez. 1990. "Structural Crises of Adjustment, Business Cycles and Investment Behaviour." In *Technical Change and Economic Theory*. Edited by G. Dosi. London: Francis Pinter.

Gaglio, Connie. 2004. "The Role of Mental Simulations and Counterfactual Thinking in the Opportunity Identification Process." *Entrepreneurship Theory and Practice* 28 (6): 533–52.

Gamson, William A. 1990 *The Strategy of Social Protest*. 2nd ed. Belmont, CA: Wadsworth Publishing Company.

Gamson, William A. 1995. "Constructing Social Protest." In *Social Movements and Cultures*. Edited by H. Johnston and B. Klandermans. Minneapolis, MN: University of Minnesota Press.

Gartner, William B. 1988. "'Who Is an Entrepreneur?' Is the Wrong Question." *American Journal of Small Business* 12 (4): 11–32.

Garud, Raghu, and Peter Karnoe. 2003. "Bricolage Versus Breakthrough: Distributed and Embedded Agency in Technology Entrepreneurship." *Research Policy* 32 (2): 277–300.

Garud, Raghu, Sanjay Jain, and Arun Kumaraswamy. 2002. "Institutional Entrepreneurship in the Sponsorship of Common Technological Standards: The Case of Sun Microsystems and Java." *Academy of Management Journal* 45 (1): 196.

Garud, Raghu, and Peter Karnoe, eds. 2001. *Path Dependence and Creation*. Lawrence Erlbaum Associates Inc.

Gaskell, George, Toby Ten Eyck, Jonathan Jackson, and Giuseppe Veltri. 2004. "Public Attitude to Nanotech in Europe and the United States." *Nature Materials* 3 (8): 496.

Gavelin, Karin, Richard Wilson, and Richard Doubleday. 2007. *Democratic Technologies? The Final Report of the Nanotechnology Engagement Group (NEG)*. London: Involve.

Gillis, Justin , and Jonathan Krim. 2004. "If It's Nano, It's BIG: Investors Are Building Mountains Out of Tiny Tech." *Washington Post* (February 22): F-01.

Godard, Olivier, Claude Henry, Patrick Lagadec, and Erwann Michel-Kerjan. 2002. *Traité des nouveaux risques. Précaution, risque, assurance*. Paris: Gallimard.

Government Printing Office. 2009. United States Code. Government Printing Office.

Greenpeace. 2005. *Annual Report*. Cited March 29, 2007. http://www.greenpeace.org/raw/content/international/press/reports/annual-report-2005.pdf.

Greenwood, Royston, and Roy Suddaby. 2006. "Institutional Entrepreneurship in Mature Fields: The Big Five Accounting Firms." *Academy of Management Journal* 49 (1): 27–48.

Greenwood, Royston, Christine Oliver, Roy Suddaby, and Kerstin Sahlin-Andersson, eds. 2008. *The Sage Handbook of Organizational Institutionalism*. Thousand Oaks, CA: Sage Publications.

Hanson, Neil. 2002. *The Dreadful Judgement: The True Story of the Great Fire of London*. London: Corgi Books.

Hardon, Jaap, Camila Montecinos, and Tim Roberts. 2005. *ETC Group External Review Report*.

Hitchcock, Darcy, and Marsha Willard. 2006. *The Business Guide to Sustainability: Practical Strategies and Tools for Organizations*. Oxford: Earthscan Publications.

Hoang, Ha, and Bostjan Antoncic. 2003. "Network-Based Research in Entrepreneurship: A Critical Review." *Journal of Business Venturing* 18 (2): 165–87.

Hodge, Graeme, Diana Bowman, and Karinne Ludlow, eds. 2009. *New Global Frontiers in Regulation: The Age of Nanotechnology*. Cheltenham, UK/Northampton, MA: Edward Elgar Pub.

Hoffman, Andrew J. 1999. "Institutional Evolution and Change: Environmentalism and the US Chemical Industry." *Academy of Management Journal* 42: 351–71.

Hopkin, Paul. 2010. *Fundamentals of Risk Management: Understanding, Evaluating and Implementing Effective Risk Management*. London: Kogan Page.

Hullmann, Angela, and European Commission. 2006. *The Economic Development of Nanotechnology—An Indicators Based Analysis*.

Hunt, Geoffrey, Michael Mehta, and eds. 2008. *Nanotechnology: Risk, Ethics and Law*. Oxford: Earthscan Publications.

Institute of Nanotechnology. 2004. *Nanotechnology in Europe*. Stirling, Scotland: Institute of Nanotechnology.

International Organization for Standardization. 2007. "Nanotechnologies." *ISO Focus* 4 (4).

Irwin, Alan , Kevin Jones, and Jack Stilgoe. 2006. *The Received Wisdom—Opening Up Expert Advice*. London: DEMOS.

IUCN, UNEP, and WWF. 1980. *World Conservation Strategy: Living Resource Conservation for Sustainable Development*. Gland, Switzerland: International Union for Conservation of Nature and Natural Resources.

IUCN, UNEP, and WWF. 1991. *Caring for the Earth: A Strategy for Sustainable Living.* Gland, Switzerland: International Union for Conservation of Nature and Natural Resources.

Jaffe, Adam, Richard Newell, and Robert Stavins. 2005. "A Tale of Two Market Failures: Technology and Environmental Policy." *Ecological Economics* 54 (2/3): 164.

Jepperson, Ronald L. 1991. "Institutions, Institutional Effects and Institutionalism." In *The New Institutionalism in Organizational Analysis.* Edited by Walter W. Powell and Paul J. DiMaggio. Chicago: University of Chicago Press.

Johnson, Dave. 2010. "Determining Nanotechnology Health Effects." *Industrial Safety & Hygiene News* 44 (1): 27–28.

Johnson, Branden, and Vincent Covello, eds. 1987. *The Social and Cultural Construction of Risk, Technology, Risk and Society.* Dordrecht, Netherlands: D. Reidel Publishing Company.

Kagan, Robert. 2004. *Of Paradise and Power: America and Europe in the New World Order.* New York: Vintage.

Kallinikos, Jannis. 2003. "Work, Human Agency and Organizational Forms: An Anatomy of Fragmentation." In *Organization Studies.*

Karra, Neri, Paul Tracey, and Nelson Phillips. 2006. "Altruism and Agency in the Family Firm: Exploring the Role of Family, Kinship, and Ethnicity." *Entrepreneurship Theory and Practice* 30 (6): 861–77.

Knol, W. H. C. 2004. "Nanotechnology and Business Opportunities: Scenarios as Awareness Instrument." Paper read at High Technology Small Firms 12th Annual International Conference, Enschede, the Netherlands, May 24–25.

Kohler, Josef. 2008. *Philosophy of Law.* Translated by A. Albrecht. Whitefish, MT: Kessinger Publishing.

Kotkin, Joel. 1994. *Tribes: How Race, Religion and Identity Determine Success in the New Global Economy.* New York: Random House.

Kuhn, Thomas S. 1962. *The Structure of Scientific Revolutions.* Chicago: University of Chicago Press.

Laas, Kelly , and Vivian Weil. 2008. "Taking a Proactive Approach towards Responsibility: Indications of Nano Policy-Making Around the World." *NBER Working Papers* 22.

Larédo, Philippe, Carole Rieu, Lionel Villard, Bernard Kahane, Aurélie Delemarle, Corineet Genet, and Vincent Mangematin. 2009. « Emergence des nanotechnologies: Vers un nouveau 'modèle industriel'? » . In *Recherche et enseignement supérieur face à l' internationalisation : France, Suisse et Union européenne.* Edited by Philippe Laredo, J.-P. Leresche, and K. Weber. PPUR Presses Polytechniques et Universitaires Romandes.

Latour, Bruno. 1988. *Science in Action: How to Follow Scientists and Engineers through Society.* Cambridge, MA: Harvard University Press.

Latour, Bruno, Steve Woolgar, and Jonas Salk, eds. 1986. *Laboratory Life.* Princeton, NJ: Princeton University Press.

Lawrence, Thomas B., and Roy Suddaby. 2006. "Institutions and Institutional Work." In *The Sage Handbook of Organization Studies.* Edited by Stewart R. Clegg, Cynthia Hardy, Thomas Lawrence, and Walter R. Nord. London: Sage Publications.

Lawrence, Thomas B., Cynthia Hardy, and Nelson Phillips. 2002. "Institutional Effects of Interorganizational Collaboration: The Emergence of Proto-Institutions. *Academy of Management Journal* 45 (1): 281–90.

Lawrence, Thomas B., Eric A. Morse, and Sally W Fowler. 2005. "Managing Your Portfolio of Connections." *MIT Sloan Management Review* 46 (2): 59–65.

Lee, Julian, and Kigali Rwanda. 2006. Global nanotechnology advocacy by NGOs. edited by CASIN. Geneva: Centre for Applied Studies in International Negotiations.

Lemley, Mark. 2005. Patenting nanotechnology. *Stanford Law Review* 58 (Nov 2005):36.

Lieberman, Marvin B., and David B. Montgomery. 1988. "First-mover advantages." *Strategic Management Journal* 9: 41–58.

Lieberman, Marvin B., and David B. Montgomery. 1998. "First-Mover (Dis) Advantages: Retrospective and Link with the Resource-Based View. *Strategic Management Journal*, 19(12): 1111–25.

Lloyd's. 2007. "Nanotechnology, Recent Developments, Risks and Opportunities." *Lloyd's Emerging Risks Team Report.*

Louisot, Jean-Paul, and Sophie Gaultier-Gaillard. 2007. *Diagnostic des risques: Identifier, analyser et cartographier les vulnérabilités.* AFNOR.

Lounsbury, Michael. 2007. "New Practice Creation: An Institutional Perspective on Innovation." *Organization Studies* 28 (07): 993–1012.

Lounsbury, Michael, and Edward J. Carberry. 2005. "From King to Court Jester? Weber's Fall From Grace in Organizational Theory." *Organization Studies* 26 (4): 501–25.

Lux Research Inc. 2004. *Sizing Nanotechnology's Value Chain.* New York: Lux Research Inc.

Lux Research Inc. 2007. *The Nanotech Report, 5th edition.* Vol. 1 & 2. New York: Lux Research Inc.

Lux Research Inc., and Matthew M. Nordan. 2005. "Nanotechnology: Where Does the U.S. Stand?" Testimony before the Research Subcommittee of the U.S. House Committee on Science, Hearing on Nanotechnology, June 29.

Maguire, Steve. 2004. "The Co-Evolution of Technology and Discourse: A Study of Substitution Processes for the Insecticide DDT." *Organization Studies* 25 (1): 113.

Maguire, Steve, and Cynthia Hardy. 2006. "The Emergence of New Global Institutions: A Discursive Perspective." *Organization Studies* 27 (1): 7–29.

Maguire, Steve, Cynthia Hardy, and Thomas B. Lawrence. 2004. "Institutional Entrepreneurship in Emerging Fields: HIV/AIDS Treatment Advocacy in Canada." *Academy of Management Journal* 47 (5): 657–79.

Malthus, Thomas R. 1999. *An Essay on the Principle of Population.* New York: Oxford University Press.

Mariani, M. 2002. "Stem Cell Legislation: An International and Comparative Discussion." *Journal of Legislation* 28: 379–411.

Meadows, Donella H., Dennis L. Meadows, Jørgen Randers, and William W. Behrens III. 1972. *The Limits to Growth: A Report for the Club of Rome's Project on the Predicament of Mankind.* 2nd ed. New York: Universe Books.

Meyer, John W., and Brian Rowan. 1977. Institutionalized Organizations: Formal Structures as Myth And Ceremony. *American Journal of Sociology* 83: 340–63.

Meyer, John W., and W. Richard Scott. 1983. *Organizational Environments: Rituals and Rationality.* Beverly Hills: Sage.

Mooney, Pat. 2001. "The ETC Century: Erosion, Technological Transformation and Corporate Concentration in the 21st Century." In *Development Dialogue.* Dag Hammarskjöld Foundation, RAFI (Rural Advancement Foundation International) news release, Winnipeg.

Mooney, Pat. 2003. "The Big Down: From Genomes to Atoms." News release from the ETC Group. Montreal: ETC.

Mooney, Pat. 2007. "Gambling with Gaia." In *ETC Group Communiqué # 90.* ETC Group.

MRG Multimedia Research Group Inc., and Fuji-Keizai, eds. 2004. *U.S. Market & Industry Nanotechnology R&D and Marketing 2004: A Guide to Application, Vendors, Technology Strategies, Product Directions & Market Development and Market Focus.* Fuji-Keizai USA and MRG, Inc.

Nanoforum. 2005. "European Nanotechnology Infrastructure and Networks." Nanoforum. http://www.nanoforum.org/nf06~modul~loadin~folder~1169~sen t~~step~~.html?.

Nanoforum, Ineke Malsch, and Mireille Oud. 2004. "Outcome of the Open Consultation on the European Strategy for Nanotechnology." NanoForum, sponsored by the European Commission under the Fifth Framework Programme.

Nanoposts 2010 Market research report. "The Global Market for Carbon Nanotubes to 2015, 2nd Ed." http://www.reportsnreports.com

NNI. 2007. The National Nanotechnology Initiative: Research and Development Leading to a Revolution in Technology and Industry. Supplement to the president's FY 2007 budget. NNI.

Nordmann, Alfred. 2007. "NanoEthics Past and Present—New Materials, Miniaturization and the Recurrent Promises of the Technological Fix." Presentation for the Office for Interdisciplinary Nanotechnology Studies, Amsterdam, February 2007. http://www.nanoOffice.eu.

North, Douglass C. 1990. *Institutions, Institutional Change and Economic Performance.* Cambridge: Cambridge University Press.

North Carolina Biotechnology Center. 2001. "Proceedings from the Workshop on Nanoscience for the Soldier," Durham, North Carolina, February 8–9.

Noymer, Andrew. 2002. "The March of Dimes." *American Journal of Public Health* 92 (2): 158.

Oberdorster, E. 2004. "Manufactured Nanomaterials (Fullerenes, C60) Induce Oxidative Stress in the Brain of Juvenile Largemouth Bass." *Environmental Health Perspectives*, 112(10): 1058–1062.

Oden, Svante. 1968. "The Acidification of Air Precipitation and its Consequences in Natural Environment." *Ecology Committee Bulletin*, Swedish Sciences Research Council, Stockholm. Arlington: VA: Translation Consultants Ltd., 1.

OECD. 1998. *Fostering Entrepreneurship.* Organisation for Economic Cooperation and Development Report Number 92–64–16139–2.

OECD. 2009. "Nanotechnology: An Overview Based on Indicators and Statistics." STI working paper 2009/7. Directorate for Science, Technology and Industry (DSTI/DOC 2009[7]): 1–112.

OECD, and Tim Davis. 2007. "Entrepreneurship Indicators Project: Developing Comparable Measures of Entrepreneurship and the Factors That Impede It." *OECD Statistics Directorate.*

Pape, Wolfgang. 1999. "Socio-Cultural Differences and International Competition Law." *European Law Journal* 5 (4): 438–60.

Patterson, K.D., and G.F. Pyle. 1991. "The Geography and Mortality of the 1918 Influenza Pandemic." *Bulletin of the History of Medicine* 65 (1): 4–21.

Paull, Robert, Josh Wolfe, Peter Hébert, and Michael Sinkula. 2003. "Investing in Nanotechnology." *Nature Biotechnology* 21 (10): 1144–47.

PCAST. 2010. *Report to the President and Congress on the Third Assessment of the National Nanotechnology Initiative.* Executive Office of the President, President's Council of Advisors on Science and Technology, Washington, D.C., March 12.

PCAST. 2005. *The National Nanotechnology Initiative at Five Years: Assessment and Recommendations of the National Nanotechnology Advisory Panel.* Office of Science and Technology Policy, Executive Office of the President, President's Council of Advisors on Science and Technology, Washington, D.C.

Phelps, Trudy A. 2006. "Introduction to European Standardization of Nanotechnologies (CEN/TC 352)." Presented at European Nanotechnology Trade Alliance Nanotechnology Policy and Planning Meeting, Brussels.

Phillips, Nelson, and John L. Brown. 1993. "Analyzing Communication In and Around Organizations: A Critical Hermeneutic Approach." *Academy of Management Journal* 36 (6): 1547.

Phillips, Nelson, and Namrata Malhotra. 2008. "Taking Social Construction Seriously: Extending the Discursive Approach In Institutional Theory." In *Sage Handbook of Organizational Institutionalism*. Edited by Royston Greenwood, Christine Oliver, Roy Suddaby, and Kerstin Sahlin-Andersson. Thousand Oaks, CA: Sage.

Phillips, Nelson, Thomas B. Lawrence, and Cynthia Hardy. 2004. Discourse and Institutions." *The Academy of Management Review* 29 (4): 635–52.

Pinchot, Gifford. (1985) 1986. *Intrapreneuring*. New York: HarperCollins. Reprint edition, New York: Harper & Row. Citations refer to the Harper & Row edition.

Pinchot, Gifford, and R. Pellman. 1999. *Intrapreneuring in Action: A Handbook for Business Innovation*. San Francisco, CA: Berret-Koehler Publishers.

Pitman, Simon. 2005. "U.S. Authorities Aim for Voluntary Nanotechnology Regulation." *The Scientist*, August 9.

Popper, Karl R. (1934) 2002. *The Logic of Scientific Discovery*. London: Routledge. Original German edition: *Logik der Forschung*, Tuebingen, Germany: Mohr Siebeck.

Porter, Michael E. 1990. *The Competitive Advantage of Nations*. New York: Free Press.

Porter, Michael E. 1998. *Competitive Advantage: Creating and Sustaining Superior Performance*. New York: Free Press.

Potoski, Matthew, and Aseem Prakash. 2004. "The Regulation Dilemma: Cooperation and Conflict in Environmental Governance." *Public Administration Review* 64 (2): 152–63.

Raje, Jaideep. 2007. "Managing Risks Associated with Nanotech Occupational Health and Safety." Presented at Nanotechnology and Occupational Health and Safety Conference, University of California, Santa Barbara.

Rao, Hayagreeva, Calvin Morrill, and Mayer N. Zald. 2000. "Power Plays: How Social Movements and Collective Action Create New Organizational Forms." *Research in Organizational Behavior* 22: 239–82.

Reynolds, Glenn Harlan. 2001. "Environmental Regulation of Nanotechnology: Some Preliminary Observations." *Environmental Law Reporter* 31 (6): 10681–88.

Reynolds, Glenn Harlan. . 2002. Forward to the Future: Nanotechnology and Regulatory Policy. San Francisco: Pacific Research Institute, 1.

Rigaud, Fabrice. 2003. « Micro et nanotechnologies: regard sur un nouveau marché high tech en Ile-de-France. » Edited by CROCIS: Centre Régional d'Observation du Commerce, de l'Industrie et des Services.

Rios, Michel. 2008. »La traduction du paradigme du développement durable: de l'appropriation des intentions stratégiques aux actes opérationnels en RSE. » *RSE Paradigm21*, juillet.

Roco, Mihail C. 2001. "Nanoparticles and Nanotechnology Research." *Journal of Nanoparticle Research* 1: 1–6.

Roco, Mihail C. 2006. "National Nanotechnology Initiative: Past, Present, Future." In *Handbook of Nanoscience, Engineering, and Technology*, 2nd ed. Edited by W. A. I. Goddard, D. W. Brenner, S. E. Lyshevski, and G. J. Iafrate. Boca Raton,FL: CRC Press.

Roco, Mihail C., and William S. Bainbridge. 2001. *Societal Implications of Nanoscience and Nanotechnology*. NSET workshop report. Arlington, VA: NSF.

Roco, Mihail C., and William S. Bainbridge, eds. 2003. *Converging Technologies for Improving Human Performance*. Dordrecht, Netherlands: Kluwer.

Romanelli, Elaine. 1991. "The Evolution of New Organizational Forms." *Annual Review of Sociology* 17 (1): 87.

Royal Society, Insight Investment, and Nanotechnology Industries Association (NIA). 2008. The Responsible Nanocode.

Rubin, Herbert J., and Irene S. Rubin. 1995. *Qualitative Interviewing: The Art of Hearing Data*. Thousand Oaks, CA: Sage.
Ruscio, Kenneth P. 1994. "Policy Culture: the Case of Science Policy in the United States". *Science, Technology and Human Values* 19(2): 205–22.
Salamanca-Buentello, F., D. L. Persad, E. B. Court, D. K. Martin, and A. S. Daar. 2005. "Nanotechnology and the Developing World." *PLoS Med* 2 (4): e97.
SCEP (Study of Critical Environmental Problems), and William H. Matthews, eds. 1970. *Man's Impact on the Global Environment: Assessment and Recommendations for Action*. Cambridge, MA: MIT Press.
Schenker, Jennifer L. 2003. "Israel Seeks Nanotech." *International Herald Tribune*, September 29.
Schumpeter, Joseph, A. 1934. *The Theory of Economic Development*. Cambridge, MA: Harvard University Press.
Schumpeter, Joseph, A. 1939. *Business Cycles: A Theoretical, Historical, and Statistical Analysis of the Capitalist Process*. New York: McGraw Hill.
Schumpeter, Joseph, A. 1947. *Capitalism, Socialism and Democracy*, 2nd ed. London: George Allen & Unwin.
Scott, Richard. 1995. *Institutions and Organizations*. Thousand Oaks, CA: Sage.
Seo, Myeong-Gu., and Douglas E. W. Creed. 2002. "Institutional Contradictions, Praxis, and Institutional Change: A Dialectical Perspective." *The Academy of Management Review* 27 (2): 222–47.
Shane, Scott. 1996. "Explaining Variation in Rates of Entrepreneurship in the United States: 1899–1988." *Journal of Management* 22 (5): 747–81.
Shane, Scott. 2003. *A General Theory of Entrepreneurship*. Edited by S. Venkataraman. Cheltenham, UK: Edward Elgar Publishing.
Sharma, Pramodita, and James Chrisman. 1999. "Toward a Reconciliation of the Definitional Issues in the Field of Corporate Entrepreneurship." *Entrepreneurship Theory and Practice* 23 (3): 11–28.
Singh, Kshitij Aditeya. 2006. "Risk Governance in Nanotechnology." Institute of Nanotechnology.
Singh, Kshitij Aditeya, and Mark Morrison. 2007. Intellectual Property in the Nanotechnology Economy. Institute of Nanotechnology.
Smalley, Richard. 2001. "Of Chemistry, Love and Nanobots." *Scientific American* 285: 76–77.
SMIC, ed. 1971. *Inadvertent Climate Modification*. Cambridge, MA: MIT Press.
Stiglitz, Joseph E. 1998. "Towards a New Paradigm for Development: Strategies, Policies and Processes." Raúl Prebisch lecture at UNCTAD, Geneva.
Stuart, Candace. 2003. Watchdogs say stop nanotech, start worldwide dialogue. *Small Times*, 31 January 2003.
Suddaby, Roy, and Royston Greenwood. 2005. "Rhetorical Strategies of Legitimacy." *Administrative Science Quarterly* 50 (1): 35–67.
Swiss Re, and Annabelle Hett. *Nanotechnology: Small Matter, Many Unknowns*. Swiss Re 2004. Cited May 2004. http://www.swissre.com/INTERNET/pwsfilpr.nsf/vwFilebyIDKEYLu/ULUR-5YNGET/$FILE/Publ04_Nanotech_en.pdf.
Tan, Sophie, and Gavin Evans. 2007. "Oil Trades Near Five-Week High After Drop in U.S. Inventories." Bloomberg report.
Taniguchi, Norio. 1974. "On the Basic Concept of Nano Technology." Paper read at Proc. Intl. Conf. Prod. Eng. Tokyo, Part II, Japan Society of Precision Engineering, 1974.
Taylor, Peter, Krishna Rajan, Bjorn Birgisson, and Tom Cackler. 2007. 'Report of the Workshop on Nanotechnology for Cement and Concrete.' Published by the *National Concrete Pavemnent Technology Center,* Iowa State University, IA, USA.

ten Wolde, A. 1998. *Nanotechnology: Towards a Molecular Construction Kit.* Chicago: New World Ventures.

Tocqueville, Alexis de. 1999. *De la Démocratie en Amérique.* Vol. 1. France: Garnier Flammarion/Philosophie.

Tolbert, Pamela S., and , Lynne G. Zucker. 1983. "Institutional Sources of Change in the Formal Structure of Organizations: The Diffusion of Civil Service Reform, 1880–1935." *Administrative Science Quarterly*, 28(1): 22–39

Tooby, J., and L. Cosmides. 2000. *Evolutionary Psychology: Foundational Papers.* Cambridge, MA: MIT Press.

Tracey, Paul. 2005. "Beyond Philanthropy: Community Enterprise as a Basis for Corporate Citizenship." *Journal of Business Ethics* 58 (4): 327–44.

Trzyna, Thaddeus. 1990. "Convening Thinkers and Doers: Sweden's Dag Hammarskjöld Foundation." In *The Power of Convening: Collaborative Policy Forums for Sustainable Development.* Edited by I. M. Gotelli. CIPA/IUCN.

UN. 1972. Declaration of the United Nations Conference on the Human Environment.

UNEP. 2007. "Emerging Challenges: Nanotechnology and the Environment." In *Global Environmental Outlook (GEO) Yearbook 2007.* UNEP.

UNEP. 2010. *UNEP Yearbook 2010: New Science and Developments in Our Changing Environment.* UNEP.

UNESCO. 2010. *UNESCO Science Report 2010: The Current Status Of Science around the World.*

United Nations Secretariat. 2009. Committee of Experts on the Transport of Dangerous Goods and on the Globally Harmonized System of Classification and Labelling of Chemicals, ST/SG/AC.10/C.4/2009/3. United Nations,

U.S. Chamber of Commerce. 2005. Regulatory Reform Issues, Precautionary Principle.

U.S. Congress. Act S. 189-2003. 21st Century Nanotechnology Research and Development Act. November 18.

U.S. Department of Commerce. 2003. *A Survey of the Use of Biotechnology in U.S. Industry.*

U.S. Department of Defense. 2005. Defense Nanotechnology Research and Development Programs.

U.S. submission to the April 10–14 Codex Alimentarius meeting in Paris addressing U.S. concerns over EU communication to the Codex on "precautionary principle.". 2000. Joint FAO/WHO food standards programme Codex Committee on general principles. Fifteenth Session, Paris, France, 10–14 April 2000.

Uzzi, Brian. 1997. "Social Structure and Competition in Interfirm Networks: The Paradox of Embeddedness." *Administrative Science Quarterly* 42 (1): 35–67.

Veleva, Vesela. 2007. "Time to Get Real: Closing the Gap Between Rhetoric and Reality." In *State Of Corporate Citizenship 2007.*

Vo-Dinh, Tuan, ed. 2007. *Nanotechnology in Biology and Medicine: Methods, Devices, and Applications.* Boca Raton, FL: CRC Press.

Vogel, David. 2003. "The Politics of Risk Regulation in Europe and the United States." *Yearbook of European Environmental Law* 3.

von Hippel, E. 1977. *The Sources of Innovation.* New York: Oxford University Press.

Wainwright, Nicholas B. 1953. "Philadelphia's Eighteenth-Century Fire Insurance Companies." *Transactions of the American Philosophical Society,* New Ser. 43 (1): 247–52.

Wardak, Ahson. 2003. *Nanotechnology and Regulation: A Case Study Using the Toxic Substance Control Act (TSCA).* Foresight and Governance Project of the Woodrow Wilson International Center for Scholars.

Wardak, Ahson. 2004. NGO Interaction with Federal Regulatory Agencies in the Context of Nanotechnology. Research paper, Rice University, unpublished.

Weart, Spencer R. 2003. *The Discovery of Global Warming (New Histories of Science, Technology, and Medicine)*. Cambridge, MA: Harvard University Press.

Weber, Max. 1997. *The Theory of Social and Economic Organization*. New York: Free Press.

Weber, Max, and H. H. Gerth. 1958. *From Weber: Essays in Sociology*. Oxford University Press, New York.

Wei-Skillern, Jane C., James E. Austin, Herman B. Leonard, and Howard H. Stevenson. 2007. *Entrepreneurship in the Social Sector*. Sage Publishing, Thousand Oaks, CA.

Wengraf, Tom. 2001. *Qualitative Research Interviewing*. London: Sage.

White, Michelle J. 2004. Asbestos and the Future of Mass Torts. *Journal of Economic Perspectives* 18 (2): 183–204.

White House. 2000. National Nanotechnology Initiative: Leading to the Next Industrial Revolution. Office of the press secretary.

WHO. 2007. Pandemic Influenza Preparedness: Sharing of Influenza Viruses and Access to Vaccines and Other Benefits. Sixtieth World Health Assembly.

Wick, Peter, Antoine Malek, Pius Manser, Danielle Meili, Xenia Maeder-Althaus, Liliane Diener, Pierre-Andre Diener, Andreas Zisch, Harald F. Krug, and Ursula von Mandach. 2010. "Barrier Capacity of Human Placenta for Nanosized Materials." *Environmental Health Perspectives* 118 (3): 432–36.

Wiesner, Mark. 2000. "Ecologie des molecules." *Libération* (French daily), December 2.

Wiesner, Mark, and Jean-Yves Bottero, eds. 2007. *Environmental Nanotechnology*. New York: McGraw Hill.

Wiesner, Mark, Gregory Lowry, Pedro Alvarez, Dionysios Dionysiou, and Pratim Biswas. 2006. "Assessing the Risks of Manufactured Nanomaterials." *Environmental Science and Technology* 40 (14): 4336–45.

Wright, Mike, Bart Clarysse, Philippe Mustar, and Andy Lockett. 2007. *Academic Entrepreneurship in Europe*. Cheltenham, UK: Edward Elgar Publishing.

Yli-Renko, Helena , Erkko Autio, and Harry J. Sapienza. 2001. "Social Capital, Knowledge Acquisition, and Knowledge Exploitation in Young Technology-Based Firms." *Strategic Management Journal* 22 (6–7): 587–613.

Zhu, Wenzhong 2008. "The Development, Trends and Challenges of Green Nanotech in Construction." In *Nanotech Northern Europe 2008*. Copenhagen, Denmark.

Zucker, Lynne G. 1977. The Role of Institutionalization in Cultural Persistence. *American Sociological Review* 42 (5): 726.

Zucker, Lynne G., and Michael R. Darby. 1998. "Entrepreneurs, Star Scientists and Biotechnology." *NBER Reporter, National Bureau of Economic Research*. Fall: 7–11.

Zucker, Lynne G., Michael R. Darby, Jonathan Furner, Robert C Liu, and Hongyan Ma. 2006. "Minerva Unbound: Knowledge Stocks, Knowledge Flows and New Knowledge Production." *Research Policy* 36 (6): 850–963.

Index